MW00951057

JOURNEYMAN ELECTRICIAN EXAM PREP

The Clearest Study Guide, with 16 Complete and Up-to-Date

Practice Tests, to Help You Easily Pass the Exam

PAUL DANIEL SMITH

© Copyright 2024 by Paul Daniel Smith —All rights reserved.

Our goal in writing this report was to provide you with knowledge that is both accurate and trustworthy regarding the subject matter.

No portion of this text may be copied, distributed, or transmitted in any manner, either electronically or in hard copy. This publication may not be copied or stored in any form without the express written consent of the publisher. Legally, we have to say that everything is off-limits.

Any liability, whether for negligence or otherwise, arising out of the receiving reader's use or misuse of any policies, methods, or directives included herein is solely the reader's responsibility. The publisher will not be held liable in any way, shape, or form for any direct or indirect damages or losses incurred as a result of using the material included herein.

All of the content provided here is for educational purposes only and applies to everyone. There is no agreement or warranty attached to the display of the data.

Paul Daniel Smith is not affiliated whit any official testing organization. All organization and test names are trademarks of their respective owners. All logos and brands mentioned herein are the property of their respective owners and are just for illustrative purposes.

This study guide is for general information and does not claim endorsement by any third party.

Table of Contents

Introduction

In general, applicants for an electrician license can only take an exam after submitting a completed licensing application to the state agency where they want to take the test (with any necessary supporting documents and payments). The applicant will be informed when they may schedule their test if they are accepted.

Additionally, applicants will be notified if their application needs to be completed or meet the requirements.

A journeyman electrician examination and 8,000 hours (4 years) of on-the-job training under the supervision of a master or journeyman electrician are typically requirements for obtaining a journeyman electrician license in most states. Typically, the test consists of 80 multiple-choice questions and lasts four hours.

A journeyman electrician license must be held for two years, a master electrician examination must be passed, and a candidate must have completed 12,000 hours (6 years) of on-the-job training under the guidance of a master or journeyman electrician. If one has a degree in electrical engineering, several states need this requirement. Typically, the test consists of 100 multiple-choice questions and lasts five hours.

Please be warned that the state where you are applying for your license may have different particular requirements and criteria.

You must fulfill certain requirements to get the title of a journeyman before you may start your career in that occupation. You will typically be regarded as a journeyman in most places once you have taken a test and obtained a license; however, the rules might vary by state and area.

The journeyman electrician exam evaluates a journeyperson's skill and knowledge. Journeymen are much more independent and responsible than apprentice electricians, so you must prove that you know what you're doing before you can complete electrical work independently. These tests are based mostly on electrical codes, which might change depending on your location.

However, you may anticipate that every test will reference the National Electrical Code (NEC), which is the foundation for many local electrical codes. No matter where you want to practice, the best approach to be ready for the exam is to fully understand the NEC.

Other organizations have produced specialized electrical regulations that may be included in your test in addition to the National Electrical Code, which the National Fire Protection Association governs. Among them are two more organizations' codes. One of them is (NESC), published every five years by the (IEEE). The International Code Council (ICC), which developed the International Building Code (IBC), the International Fire Code (IFC), and the International Energy Conservation Code, are the other organizations that offer regulations and information for building safety professionals (IECC).

To tailor your studying to your particular exam, you must know which of the criteria above you will be assessed in the fields in which you want certification to practice. You may ask your local or state's licensing agency or oversight board which codes you must know. Additionally, students who received their education at a local community college or trade school may be eligible to enroll in a curriculum-integrated review course that is regionally oriented.

Chapter 1:
The Journeyman Electrician Exam

The state or local municipality must require an journeyman electrician's license before allowing someone to work as a contractor. This test attests to the fact that you possess the talents required to carry out your duties to the best of your ability while keeping consumer and public safety in mind.

1.1 Who Can Be Admitted to the Examination?

Most states and municipal governments demand that those who want to become journeyman electricians have a certain amount of professional experience in the field. Some people may additionally demand that you take part in certain educational initiatives. Your region often requires a completed apprenticeship under the guidance of a Journeyman electrician for a certain period, typically a year or more.

It would be best if you got in touch with the state licensing body or the testing business that handles such activities to ensure that you have complied with all the criteria for your region. They will be able to provide you with detailed instructions on how to register for and schedule a journeyman electrician's test and what has to be done before you apply for such a license. Typically, there is a charge to take the test. Usually, it costs roughly $75.

When and Where Can I Take It?

Several times a year, testing locations are often held at a nearby testing firm. You should get a list of venues and days open in your region from your state licensing agency.

You must show up for your test at least 30 minutes early. Doing this will ensure everyone has enough time to register and be ready before the test starts. If you are late, you won't be able to take the exam that day, and you'll have to reschedule and pay another exam cost.

What items should I bring?

You may need to carry official identification with you to the exam location. It must show your written name, a picture of you, your signature, your birthdate, and an expiry date. All of the information used to register you for the test and this ID must be identical.

Generally, you can only carry your appropriate ID and a quiet, battery-powered, non-programmable calculator into the testing area. Electronic gadgets, large or loose clothes, food, and beverages will all be given a special storage space outside the examination room.

The test will often be given on a computer equipped with a mouse and keyboard. Although in certain places, you may also take the test using paper and pencil.

A soft-bound copy of the National Electrical Code can often be used as a reference during the test since most examinations are "open books."

What does the test cover?

The only way to know what will be on your exam is to get in touch with your state's licensing body or testing provider since each exam varies significantly by state. With between 80 and 100 scored multiple choice and true/false questions, the majority are nonetheless reasonably comparable.

A few things on the test need to be graded. These are often not noted as such, and you will know about them once you get your findings. Depending on the number of items on the exam, you will typically have 240 minutes to finish it.

The following are the topic areas commonly tested for:

- Plans, theories, calculations, and definitions.
- Electrical services, service items, and systems that are independently derived.
- Branch circuit calculations for electrical feeders and conductors.
- Electrical materials and wiring techniques.
- Electrical devices and equipment.
- Generators and motors.
- Electrical control mechanisms and methods of disconnection.
- Special occupations, gear, and situations.

When Will I Discover My Outcomes?

Immediately. You will get the relevant message on the computer screen after the test. You will also be given a printed copy of your official score report at the exam location. Typically, you need a score of at least 70% to pass.

1.2 Number of Questions

1. There are 80 questions on the licensure exams for Class A master electricians and Class A journey worker electricians.

2. The power-limited technician licensing test consists of 80 questions.

3. The maintenance electrician and lineman licensure exams each include 70 and 50 questions, respectively.

4. There are 25 questions on the licensing test for Class B and the license for installing satellite systems.

5. Unless otherwise specified, each test question is worth the same number of points. There are no partial points; each question receives either a full score or no points.

6. Seventy percent is the cutoff for all exams.

1.3 Requirements for Journeyman Electricians by State

It would help if you looked into the test criteria for the fields you want to work in, as previously indicated. The table below details the numerous codes you must learn to pass the test offered in your state or locality. You should check with your local professional licensing board for specifics to enhance this information.

1. State: Alabama

- **Exam Required?**   Yes
- **At the State or Local Level?** State wide
- **Exam References:** NEC, IFC
- **Usable in Other States?** AK, GA, LA, TN, VA

2. State: Alaska

- **Exam Required?**   Yes
- **At the State or Local Level?** State wide

- **Exam References:** NEC, NESC
- **Usable in Other States?** AK, CO, ID, MN, MT, NE, WA, WY

3. **State:** Arizona

- **Exam Required?**   Yes
- **At the State or Local Level?** Local
- **Exam References:** NEC
- **Usable in Other States?** CA, NE, UT

4. **State:** Arkansas

- **Exam Required?**   Yes
- **At the State or Local Level?** Local
- **Exam References:** NEC
- **Usable in Other States?** AL, LA, MS, TN

5. **State:** California

- **Exam Required?**   Yes
- **At the State or Local Level?** State wide
- **Exam References:** CA Electrical code, based on NEC, and CA Fire code, based on IFC
- **Usable in Other States?** AZ, NV, UT

6. **State:** Colorado

- **Exam Required?**   Yes, the exam is open-book
- **At the State or Local Level?** State wide
- **Exam References:** NEC
- **Usable in Other States?** AK, ND,

7. **State:** Connecticut

- **Exam Required?**   Yes, the exam is open-book
- **At the State or Local Level?** State wide
- **Exam References:** Electrical Board of Occupational Licensing, similar to NEC
- **Usable in Other States?** No

8. **State:** Delaware

- **Exam Required?**   Yes, the exam is open-book
- **At the State or Local Level?** State wide
- **Exam References:** NEC

- **Usable in Other States?** No

9. State: District of Columbia

- **Exam Required?** Yes, the Exam is open-book
- **At the State or Local Level?** State wide
- **Exam References:** NEC
- **Usable in Other States?** No

10. State: Florida

- **Exam Required?** Yes
- **At the State or Local Level?** Local
- **Exam References:** NEC
- **Usable in Other States?** AL, GA, NC

11. State: Georgia

- **Exam Required?** No
- **At the State or Local Level?** No
- **Exam References:** No
- **Usable in Other States?** AL, FL, NC, SC

12. State: Hawaii

- **Exam Required?** Yes
- **At the State or Local Level?** State wide
- **Exam References:** NEC
- **Usable in Other States?** No

13. State: Idaho

- **Exam Required?** Yes
- **At the State or Local Level?** State wide
- **Exam References:** NEC
- **Usable in Other States?** AK, CO, MN, MT, NE, UT, WA, WY

14. State: Illinois

- **Exam Required?** Yes
- **At the State or Local Level?** Local
- **Exam References:** NEC
- **Usable in Other States?** No

15. State: Indiana

- **Exam Required?**   Yes
- **At the State or Local Level?** Local
- **Exam References:** NEC
- **Usable in Other States?** No

16. State: Iowa

- **Exam Required?**   Yes
- **At the State or Local Level?** Local
- **Exam References:** NEC, IBC
- **Usable in Other States?** No

17. State: Kansas

- **Exam Required?**   Yes
- **At the State or Local Level?** Local
- **Exam References:** Kansas City Fire Prevention Code, NEC
- **Usable in Other States?** No

18. State: Kentucky

- **Exam Required?**   Yes
- **At the State or Local Level?** State wide
- **Exam References:** NEC
- **Usable in Other States?** OH

19. State: Louisiana

- **Exam Required?**   Yes
- **At the State or Local Level?** State wide
- **Exam References:** NEC
- **Usable in Other States?** AL, UT

20. State: Maine

- **Exam Required?**   Yes
- **At the State or Local Level?** State wide
- **Exam References:** NEC
- **Usable in Other States?** MA, NH, OR, VT

21. State: Maryland

- **Exam Required?** Yes
- **At the State or Local Level?** Local
- **Exam References:** (MBPS), (MBRC)
- **Usable in Other States?** DE, VA

22. State: Massachusetts

- **Exam Required?** Yes
- **At the State or Local Level?** State wide
- **Exam References:** Massachusetts Electrical Code, based on NEC
- **Usable in Other States?** ME, NH, OR, VT, WA

23. State: Michigan

- **Exam Required?** Yes
- **At the State or Local Level?** State wide
- **Exam References:** Michigan Electrical Code, based on NEC
- **Usable in Other States?** No

24. State: Minnesota

- **Exam Required?** Yes
- **At the State or Local Level?** State wide
- **Exam References:** Minnesota State Building Code, NEC, NESC
- **Usable in Other States?** AL, CO, ID, OR, SD, UT, WA, WY

25. State: Mississippi

- **Exam Required?** Yes
- **At the State or Local Level?** Local
- **Exam References:** It is based on the code of each jurisdiction
- **Usable in Other States?** AL, AK, LA, TN

26. State: Missouri

- **Exam Required?** Yes
- **At the State or Local Level?** Local
- **Exam References:** NEC
- **Usable in Other States?** No

27. State: Montana

- **Exam Required?** Yes
- **At the State or Local Level?** State wide
- **Exam References:** NEC
- **Usable in Other States?** AL, AK, CO, ID, UT, WA, WY

28. State: Nebraska

- **Exam Required?** Yes
- **At the State or Local Level?** State wide
- **Exam References:** NEC
- **Usable in Other States?** AL, AK, UT, WA, WY

29. State: Nevada

- **Exam Required?** Yes
- **At the State or Local Level?** Local
- **Exam References:** NEC
- **Usable in Other States?** No

30. State: New Hampshire

- **Exam Required?** Yes
- **At the State or Local Level?** Local
- **Exam References:** NEC, NESC
- **Usable in Other States?** ME, MA, VT

31. State: New Jersey

- **Exam Required?** No
- **At the State or Local Level?** No
- **Exam References:** No
- **Usable in Other States?** DE

32. State: New Mexico

- **Exam Required?** Yes
- **At the State or Local Level?** State wide
- **Exam References:** NEC, New Mexico Electrical Code
- **Usable in Other States?** No

33. State: New York

- **Exam Required?** Yes
- **At the State or Local Level?** Local
- **Exam References:** NEC
- **Usable in Other States?** No

34. State: North Carolina

- **Exam Required?** Yes
- **At the State or Local Level?** Local
- **Exam References:** NEC
- **Usable in Other States?** CA, NE, UT

35. State: Arizona

- **Exam Required?** Yes
- **At the State or Local Level?** Local
- **Exam References:** NEC, Building Code Council
- **Usable in Other States?** MS, SC, VA, WV

36. State: North Dakota

- **Exam Required?** Yes
- **At the State or Local Level?** State wide
- **Exam References:** NEC
- **Usable in Other States?** AK, AR, CO, WY

37. State: Ohio

- **Exam Required?** Yes
- **At the State or Local Level?** Local
- **Exam References:** NEC
- **Usable in Other States?** KY, WV

38. State: Oklahoma

- **Exam Required?** Yes
- **At the State or Local Level?** State wide
- **Exam References:** NEC
- **Usable in Other States?** AK, AR, CO, MN, NM, ND, SD, UT, WA, WY

39. State: Oregon

- **Exam Required?** Yes
- **At the State or Local Level?** State wide
- **Exam References:** EC, OESC
- **Usable in Other States?** AK, MT, UT, WY

40. State: Pennsylvania

- **Exam Required?** Yes
- **At the State or Local Level?** Local
- **Exam References:** NEC
- **Reciprocity With Other States:** CT

41. State: Puerto Rico

- **Exam Required?** Yes
- **At the State or Local Level?** State wide
- **Exam References:** NEC
- **Usable in Other States?** No

42. State: Rhode Island

- **Exam Required?** Yes
- **At the State or Local Level?** State wide
- **Exam References:** NEC, Rhode Island State Building Code
- **Usable in Other States?** No

43. State: South Carolina

- **Exam Required?** No
- **At the State or Local Level?** State wide
- **Exam References:** NEC
- **Usable in Other States?** AL, GA, MS, NC, TN, TX, UT

44. State: South Dakota

- **Exam Required?** Yes
- **At the State or Local Level?** State wide
- **Exam References:** NEC, NESC
- **Usable in Other States?** AL, WY

45. State: Tennessee

- **Exam Required?**   Yes
- **At the State or Local Level?** Local
- **Exam References:** State Fire Marshall Regulation
- **Usable in Other States?** AL, MI

46. State: Texas

- **Exam Required?**   Yes
- **At the State or Local Level?** State wide
- **Exam References:** NEC
- **Usable in Other States?** No

47. State: Utah

- **Exam Required?**   Yes
- **At the State or Local Level?** State wide
- **Exam References:** NEC
- **Usable in Other States?** AK, AR, WY

48. State: Vermont

- **Exam Required?**   Yes
- **At the State or Local Level?** State wide
- **Exam References:** NEC
- **Usable in Other States?** ME, NH

49. State: Virginia

- **Exam Required?**   Yes
- **At the State or Local Level?** State wide
- **Exam References:** NEC, ICC
- **Usable in Other States?** MD, NC

50. State: Washington

- **Exam Required?**   Yes
- **At the State or Local Level?** State wide
- **Exam References:** NEC
- **Usable in Other States?** AK, AR, CO, ID, MT, SD, UT, WY

51. State: West Virginia

- **Exam Required?**   Yes
- **At the State or Local Level?** State wide
- **Exam References:** NEC
- **Usable in Other States?** NC, OH, VA

52. State: Wisconsin

- **Exam Required?**   Yes
- **At the State or Local Level?** State wide
- **Exam References:** NEC
- **Usable in Other States?** No

Chapter 2:
The Electrical System

Because of the potential for fires and explosions to result from improper handling and installation of electrical wiring, specific standards must be adhered to in the selection of materials, the quality of the work, and the safety precautions that must be taken. The National Electrical Code was formulated to standardize, organize, and simplify these laws and offer a reliable reference for electrical building. Since it was first published in 1897, the National Electrical Code (NEC) has been updated multiple times to consider new technological developments, advancements in building materials, and rising dangers posed by fire. It is the result of a coordinated effort by electrical engineers, manufacturers of electrical equipment, insurance underwriters, firefighters, and other specialists from different parts of the country.

The National Fire Protection Association (NFPA), located at Battery March Park, Quincy, Massachusetts 02269, is currently in charge of publishing the NEC. It contains specific laws and regulations intended to assist in protecting persons and property in the real world against the dangers associated with electric-related hazards.

Even though the NEC makes it clear that "This Code is not designed as a design specification nor as an instruction manual for unskilled employees," it nevertheless acts as a reliable basis upon which the study of electrical design and installation can be constructed. This Code contains precautions regarded as necessary for safety, which is probably why the NEC conducts its safety research.

Following these rules will result in a safe installation, but it might not be optimal in terms of efficiency, convenience, or capacity for the electrical loads currently being used or intended to be used in the future.

On the other hand, the NEC has established itself as the standard for the industry and is cited on most licensure tests for electricians and electrical contractors. As a result, everyone who works with electricity should acquire a copy, ensure it is readily available at all times, and consult it frequently.

2.1 What is a Power Grid?

The network of interconnected systems referred to as the "electrical power grid" is responsible for transporting electricity from the generators to the power users. A power grid or supply is another name for this type of infrastructure. Transmission lines, power plants, and distribution networks constitute the three main elements that make up an electric grid.

Stations that generate electricity are typically positioned in areas where doing so is not only feasible but also advantageous, such as close to sources of fuel, dams, or sources of renewable energy. Because of this, they are typically located in secluded areas, far away from the next major city. This makes a lot of sense because transporting electricity over long distances is much more cost-effective than transporting gas or oil. It is possible to construct wind power plants offshore to collect a greater quantity of energy from the wind. Hydroelectric power plants, on the other hand, need to be located close to suitable dam sites. It is necessary to have a transmission infrastructure that spans great distances to bring electricity from the generators to the populated areas. In addition, a distribution system is essential to ensure that the energy is delivered to each user at the appropriate voltage.

Types of Power Grid

Because the power plant's location is so close to its source of fuel, the grid's transportation expenses are kept to a minimum. However, you are quite a distance away from any populated area. Before the power is sent out to clients, it goes via a step-down transformer in the substation, which brings the high-voltage electricity's voltage down to a more manageable level. There are primarily two types of components that make up the electrical grid. These objects are as follows:

Regional Grid:

The regional grid is what we call the network created when all of the transmission systems in a particular region are connected through the use of lines.

National Grid:

It is the result of connecting several different regional grids.

2.2 Cause for the Existence of an Interconnection

The interconnection of the grid allows for the most efficient use of the available electricity resources and provides a high level of supply security. Because of this, the system is both cost-effective and reliable. The power stations that produce electricity are interconnected, so there is less of a need for reserve generation capacity in each region.

If a zone has a rapid rise in load or a loss of production, it will borrow from the next interconnected region. However, to ensure the smooth operation of the network's interconnections, a certain amount of producing capacity, also referred to as distributed generators, is necessary. The revolving reserve consists of a generator operating at its usual speed and prepared to give power immediately.

2.3 Types of Interconnections

The connectivity between networks can primarily be broken down into two categories: HVDC link and HVAC link, respectively.

High Voltage Alternating Current Interconnection

Interconnection of HVAC is also known as high-voltage alternating current. An AC link is what connects the two different air conditioning systems that are part of an HVAC link. The frequency control on each of the two systems must be sufficiently close to one another for the AC system to be interconnected.

In a system that operates at 50 hertz, the frequency must be somewhere in the range of 48.5 to 51.5 hertz. The term "synchronous interconnection" is used to refer to this type of connection. The AC link provides a stable connection between two AC systems that will be joined. However, several restrictions are associated with the AC connector.

The following difficulties have been encountered throughout the process of interconnecting an AC system.

- o The term "synchronous tie" refers to the connection that allows the two AC networks to communicate with one another. Both of these systems will experience frequency disturbances as a result of the interaction between them.

- o The fluctuations in power in one system affect the other system. Large power swings in a single system can cause frequent tripping, which can lead to serious faults in that system. This error brings down the entire interconnected structure and leads it to fail.

o If one AC system is connected to another AC system using an AC tie line, there will be a rise in the probability that a failure will occur in both of the systems. This is because the additional parallel line lowers the equivalent reactance of the system that is coupled. If both of the AC systems are linked to the fault line, then the fault level of both of the AC systems will continue to be the same.

High-Voltage Direct Current Connections

The DC link, also known as a DC tie, is a technology that provides a relatively loose coupling between two AC systems. It is also known by its acronym, DC. A direct current link connects two asynchronous direct currents (AC) systems (Asynchronous). Using DC connections between devices can be beneficial in several ways. The goods in question are those listed above.

o Because the DC interconnection method is asynchronous, the systems that will be connected need to either have the same frequency or run at different frequencies to be connected. As a result, the DC link offers the advantage of joining two AC networks that operate at different frequencies. Because of this, the system can meet the frequency requirements it has set for itself and continues to operate independently.

o By modifying the firing angle of the converters, direct high-voltage current (HVDC) links make it feasible to rapidly and precisely change the amount of power flow and the direction in which it travels. Because the flow of power may be quickly adjusted, the transient stability limit is significantly higher.

o Dampening power swings in linked AC networks can be accomplished quickly through the use of power flow modulation through the DC tie. As a direct consequence of this, the system becomes more stable.

In recent years, traditional power grids have been largely replaced by their digital counterparts, known as smart grids. Because of the smart meter and other intelligent pieces of equipment, the smart grid operates more effectively.

2.4 How Does the Power Grid Function?

The generation, distribution, and transmission of electricity are the three primary functions that make up a power grid. The following text provides elaborate explanations of these steps.

- **Power Production**

The facilities responsible for the production of energy, known as power plants, are often located in isolated areas far off from places that have a high population density. Power plants that run on thermal, nuclear, hydroelectric, solar, or wind energy are just a few of the accessible alternatives. In a power plant, power can be generated using two or more parallel 3-phase alternators or even more than that. Power plants generate electricity at voltages between 11 kilovolts and 25 kilovolts. Because of several technical limitations, there is no way to greatly boost the produced voltage.

- **Transmission**

To transport energy over greater distances, the resulting voltages are amplified to significantly higher levels before being sent on their way. Because the current declines as the voltage increases, a step-up transformer is required to compensate for this phenomenon. Boosting transmission efficiency and reducing the impact of I2R losses can be accomplished in part by increasing the voltage carried across the transmission lines. Increasing the transmission voltage makes it possible to produce a lower transmission current and, as a result, a smaller I2R loss. From 220 kilovolts and up to 765 kilovolts, transmission voltages can range everywhere. Transmission wires frequently can be seen climbing up and over tall towers when one approaches the outskirts of a city.

Power transmission often uses extremely high voltage alternating current (AC) that operates in three phases. As a result of advancements in power electronics, however, HVDC, also known as high voltage direct current, has demonstrated considerable advantages for transmission across greater distances. For this reason, we are sending power across oceans and continents utilizing direct high-voltage current (HVDC) systems. To send energy across long distances, alternating current (AC) power must first be transformed into direct high-voltage current (HVDC) at a converter station. Additionally, the utilization of HVDC cables is the sole option available at this time for connecting grids that operate at various frequencies.

- **Distribution**

Using a step-down transformer, a primary step-down substation takes electricity from the transmission system and drops the voltage by a large amount, typically on the order of 33 to 66 kilovolts, for example. The power is either delivered to substations for distribution or to major industrial users.

At the distribution substations, the power is decreased even further (say at 11kV). Power is often distributed through the use of distribution lines that are above or below ground, and these lines are typically connected in the form of a ring or a mesh network. The voltage is further decreased by distribution transformers until it reaches the utilization voltage, and secondary distribution lines are used to supply many users (120 volts or 230 volts).

Having said that, this is merely a summary of the general principles behind how a power grid operates. In contrast, the complexity of an actual electrical grid is a great deal larger than what is often believed. The transmission and distribution voltages might be very different from one another.

2.5 Electric Circuits

A path along which electric current can travel is referred to as a circuit. By traveling through a circuit, electricity can be converted into other forms of energy that can then be utilized to power lights, appliances, and other electronic devices. These alternative forms of energy can then be put to work.

For a complete understanding of the properties of electricity, it is vital to have a working knowledge of electric current and electric circuits. To put it another way, electrons, protons, and ions are all examples of charged particles that can be found moving along the path of an electric current. Because direct and alternating charging are both possible in a circuit, the names given to these two current forms reflect this fact. In a circuit that uses direct current (DC), the flow of electric charges is always consistent in one direction. Charges on the conductors of a circuit will alternate directions and pulsate at a high frequency whenever an alternating current (AC) is passed through it.

Wires, a device (such as a light bulb or motor) that uses the current, and a power source are the three components always present in an electric circuit. It is true regardless of how complex the circuit may be (such as a battery or generator). The generators of a power plant are responsible for supplying homes and other significant buildings with electricity. To ensure that a circuit is operational, all of its components must be connected. The circuit is considered closed once all of the connections have been made, at which point electricity can flow without restriction. When a conductor's connection to the circuit is severed, no current will flow across the circuit. Using a switch, we can selectively activate or deactivate the current flow in a circuit. Turning on a light switch is the only way to finish the electrical circuit. Because there is no longer anything blocking the flow of current, the light bulb can finally be turned on. The light goes out because the circuit is broken when the switch is turned off, which also results in the power to the lamp being turned off.

2.6 Different Types of Electrical Circuits

In addition to the classifications determined by the information transmitted, there are other more fundamental types of electric networks, the purpose of each of which is to carry power.

However, before discussing the many kinds of circuits, it is necessary first to define the terms "open circuit" and "closed circuit." The creation of a closed path is what differentiates a closed circuit from an open circuit. An open circuit does not provide a closed path like a closed circuit. Electricity can travel across a circuit if and only if the circuit is closed.

- **Series Circuits**

In a circuit organized in series, there is only one conductor. The power source, the cables, and the various devices are all linked along the same channel in a series circuit. Hence there are no branches in this type of circuit. The process begins with the first apparatus and moves on to the subsequent ones in order.

In a circuit organized in series, the amount of current that passes through each component remains the same. However, if more elements are added to a series circuit, the current flowing through each component will decrease as a result of the increased number of elements. When you add more lights to a series circuit, for instance, the light produced by each light will be dimmer than previously. When there are fewer lights, there is a smaller amount of electricity that can be transferred to each light. No matter how many different items are connected to the circuit, the current in a series circuit will never be "used up" because the circuit operates differently.

If one of the components in a series stops working, the complete circuit is inoperable, and the remaining components will not receive any current. Common examples of gadgets that use series circuits are torches, Christmas lights, and other similar items.

- **Parallel Circuit**

Multiple branches, or routes, exist in a parallel circuit. Each component is linked to its separate electrical circuit. The current in a parallel circuit will become divided if it comes into contact with a branch. Because only a small portion of the total current flows via each branch of the circuit, the amount of current circulating at various points in the circuit varies.

The total amount of current available to each branch of a parallel circuit remains the same as more branches (and devices) are added to the circuit. This is because the flow of current in each branch is independent of the flow of current in the other branches. If you add more lights to a parallel circuit, each one will shine just as brilliantly as it did before as long as the new lights are attached to different branches of the circuit. This is the only condition that must be met for this to be the case.

It is possible to turn on or off a device located in one branch of a parallel circuit without having any effect on the other branches. If one component in a parallel circuit fails or is unplugged, the others will continue functioning normally, even if the circuit is disrupted. Parallel circuits are used whenever more than one device needs to continue running if there is an interruption in the electrical supply.

- **Open Circuit**

An open circuit is a circuit that is not completed, meaning there is no way for current to flow back through it, so it does not have a return path. An open circuit is a type of circuit that maintains a voltage proportional to the producing source's EM field but does not experience any passage of current at all.

- **Close Circuit**

A closed circuit is the same thing as a completed circuit, which simply means that there is a way for current to flow back through the circuit.

- **Short Circuit**

It is a circuit where the current flow is in both directions and which has a point where the resistance value is zero. A finished or closed circuit that does not have any load connected to it is referred to as a short circuit. In other words, a short circuit is a type of circuit that has a voltage that tends to approach zero and a current that tends to approach infinite.

- **Series-Parallel Circuit**

A series-parallel circuit is one in which the circuit elements are linked in series in some sections and parallel in other portions. To put it another way, this hybrid circuit combines elements of series, parallel, and series-parallel configurations.

- **Star-Delta Circuit**

In these circuits, the connection of the various electrical elements is not delineated in terms of whether they are arranged in series, parallel, or series-parallel configurations. The transformation from star to delta and back to star can be used to solve problems involving star-delta circuits.

- **AC Circuit**

AC circuits are any circuit powered by an alternating current (AC) supply source of voltage. Alternators and synchronous generators are two examples of different supply sources.

- **DC Circuit**

DC circuits are any circuits that have a direct current (DC) supply source of voltage inside their design. Batteries and direct current generators are some examples of supply sources.

- **Complex Circuit**

Complex circuits can't be simplified down to a single resistor, and their components don't fit neatly into either series or parallel configurations. In this particular kind of circuit, the connections between the resistors are quite intricate.

Multiple sources of electromotive force or pure voltage are typically present in circuits that have a high degree of complexity. It is not possible to solve them by combining elements in a series or parallel.

Chapter 3:
Electrical Principles and Formulas

Every day, we use electricity, but have you ever stopped to think about how the whole thing is put together? Electric current, in its most basic form, travels along the length of a conductor in the form of free electrons, which move from one atom to the next as they do so. Think back to the science subjects you took in high school you believed you would never use again. The current will conduct electricity more efficiently if it contains a greater number of free electrons.

3.1 Basic Electric Principle

It is helpful to have a fundamental understanding of how the components of an electrical system function to alleviate the dread and anxiety connected with working with electricity. Our first topic of discussion will be the scientific study of electricity.

Atoms are the fundamental components of all that there is in the universe. Atoms have a nucleus at their center and electrons orbit around the nucleus in a circular pattern. According to certain models, one way to conceptualize it is as a central sun surrounded by planets in orbit around it (the electrons). The total number of electrons that orbit around the nucleus is what determines the atomic characteristics of an element. The simplest atom is hydrogen, which only possesses one electron, whereas the most complex atom is copper, which possesses twenty-nine electrons. Hydrogen is the simplest atom. The negative charge carried by each electron is balanced out by a positive charge present in the nucleus of the atom. The atom as a whole does not have a positive or negative charge.

- **Flow of Current**

To put it another way, when it comes to the forces exerted on electrons, which carry charge-like polarities, they tend to repel one another, whereas opposing polarities tend to attract one another. Electrons may be brought to a substance if an external positive charge has been given to it, as this charge may attract electrons. This can be related to the two forces in the same way as magnetic effects can, where similar poles repel and different poles attract each other. An electric current exists when electrons move in a steady stream from a region with a negative charge to one with a positive charge.

Electric current is composed of electrons that are continuously moving from one atom to another. These electrons are being driven in a certain direction by an outside potential, and as a consequence, they will be pulled to the positive. According to the concept of current, which defines the flow of positive charge, the current direction will be opposed to that in which the electrons are flowing. It happens because a positive charge moves faster than a negative charge. If we disregard this peculiarity, we may conceive electric current as the passage of electrons from a positive terminal to a negative terminal. This helps us understand how electric current works. Current is used as a measurement tool to determine the rate of electron flow.

The charge may be produced whenever the normal distribution of electrons to nucleons is disrupted. A positive charge is acquired by an object whenever it undergoes the process of electron loss. The addition of electrons creates a change in the charge of a substance, moving it from positive to negative.

Since electrons' overall architecture is not changed by a circuit, electrons that enter a circuit will have the same characteristics as electrons that exit the circuit. As a direct consequence of this, none of the parts of the circuit will pick up any charges. The key realization that electrical charge does not disappear is one that naturally derives from this observation. Whatever current is sent into a circuit should eventually find its way out of it. The current cannot be stopped by anyone's efforts. The current may leak out of the circuit as a result of the fault scenario (for example, there may be a short circuit to earth). At all times, the current going into the circuit has to be equivalent to the current exiting the circuit. The ability to detect current leakage is provided by Residual Current Devices thanks to this principle. When an electromagnetic force, also known as an EMF, is applied to a circuit, an electric current is produced. Because there is a current present, there is the potential for more than one conclusion to occur.

- **Chemical Effect**

When subjected to an electric current, certain substances go through chemical transformations, although metals, for the most part, remain unaffected. When dissolved in water, acid first disperses and then decomposes into its component elements. Electroplating and accumulators are two examples of processes that use the chemistry that is induced by an electric current.

- **Heating Effect**

When an electric current is passed through substances that have some resistance to the flow of electric current, the temperature of those substances will rise. Some conductors are constructed in such a way as to have a predetermined level of resistance for them to successfully use the heating effect. The elements that make up an immersion heater are a stretch of cable that is mineral-insulated and contains resistance wires in their conductor positions. An electric fire element's beating heart is made up of resistance wire wound around a heat-resistant insulating rod. A filament is a tiny resistance wire that can be found inside an electric light bulb. Because of the current, the filament eventually reaches a temperature at which it turns incandescent and emits light.

- **Magnetic Effect**

When electric current moves through a conductor, it creates a magnetic field. The polarity of the magnetic field, as well as its strength, is proportional to the magnitude and direction of the electric current. Motors use the forces developed as a result of the interplay of magnetic fields to generate motion.

3.2 Electric Units

- **Voltage**

Voltage refers to the driving force behind electricity, often known as pressure. The word electromotive force, or emf for short, is the one used to refer to this particular quantity, even though it's not very frequent. In an electrical system, the voltage, which is also measured in volts, is typically the component that remains constant. Within a circuit, voltage serves as the energy source and can originate from various locations. Within a DC system, an example of a source of continuous DC voltage would be something like a battery or a DC Power Supply Unit. A freestanding generator will supply an AC system with the necessary alternating current (AC) power.

- **Current**

When a voltage is added to a circuit, electrons will begin to flow across the surface of the circuit. The unit of measurement for current is the ampere (amperes). About six billion, billion, billion electrons pass across a conductor in one second for every ampere. When voltage is applied to a closed circuit, current will flow across the circuit (or path). The overall current in a circuit is controlled by the resistance present in the circuit.

- **Resistance**

The material itself is one of the elements that determine resistance. This is because every substance possesses a quality referred to as resistivity. Electrons in that substance must get out of their orbits around the nucleus to participate in the passage of electric current. Insulators are any materials that prevent electrons from moving freely through them, whereas conductors are any materials that make this process easy to carry out. The ohms per meter is the unit of measurement for resistivity.

In electrical circuits, resistance performs the function of a barrier to the flow of current. The amount of force that must be exerted by an electric current to move through a certain substance is referred to as the material's "resistance," and it is measured in Ohms.

The length of the material and the area of its cross-section both play a role in determining its dimensions. To get a better idea of how resistant something is:

Resistance = Resistivity of the material X Length/Cross-sectional area

The relationship between resistance and dimensions, such as length and cross-sectional area, is made abundantly evident by using the formula mentioned above.

The resistance increases with increasing length but decreases with increasing thickness. When there is a greater cross-sectional area, there will be less resistance, and vice versa.

This is crucial since different tasks require various diameters of cables to be used. Using cables that are either too long or too thin can result in an excessive amount of resistance being injected into the system.

- **Conductors**

In a conductor, the electrons are not firmly attached to the nucleus; rather, they are free to wander about from one atom to the next as they please. They rapidly leave their orbit when an external potential is supplied, which results in current flow. Conductors are defined as materials with low resistance, allowing current to flow easily through them. Silver, copper, and aluminum are examples of some of the metals that can be found in this category.

- **Insulators**

When an external potential is applied to an insulator, the electrons cannot detach from the nucleus and move onto a new orbit as they would if the insulator were not subjected to the potential. Because there will be no exchange of electrons, there will be no flow of current. An insulator refers to a material that does not conduct electricity or a substance that has a very high resistance such as rubber, polyvinyl chloride, and porcelain.

- **Energy**

A joule is a common unit of measurement for energy, which may be thought of as the capacity of an object or system to carry out work.

Example

1. A lamp can transform electrical energy into light energy.

2. A motor is an electromechanical device that can transform electrical energy into mechanical energy.

3. The energy of a battery is given by the conversion of chemical energy into electrical energy.

The water's potential energy in a highland reservoir is changed into the water's kinetic energy as it travels down the input tube and is then used to power a generator propelled by a turbine. This results in the water having electrical energy.

- **Power**

Watts are the unit of measurement of power; they represent the rate at which energy is absorbed or delivered. In the vast majority of instances, the conversion of energy does not achieve perfect efficiency, meaning that some of the input power is used to produce the desired output. The ratio of the total power wanted at the output to the total power that is entered is what the efficiency of the system indicates.

Efficiency = <u>Power Output</u> X 100%

Power Input

In every case, an additional kind of energy is generated as a side product. The majority of the time, this by-product is heated, and it is frequently important to discover techniques for getting rid of this heat by offering additional cooling methods, such as having fins or fans. If this heat is not removed, the apparatus may overheat, which might have potentially disastrous results. It is essential to understand the method by which a piece of machinery is cooled, as any factor that impedes this process can harm the machinery.

If you restrict the flow of air, it could disrupt the cooling, which would lead to an increase in temperature.

Resistance results in heat production, which leads to a loss of power. Resistance is something that conductors have, causing heat to be generated and a loss of power. For example, a loudspeaker coil possesses resistance. This resistance will result in the production of heat, but it is not required for the process of transforming electrical power into sound power. To prevent the speaker from overheating, this heat will need to be removed from the system.

3.3 Electric Power

Electric power can be defined as the rate at which work is accomplished in a circuit that uses electricity. To put it another way, the rate at which energy is transported can be thought of as the definition of electric power. The generator is responsible for producing the electric power, although the electrical batteries can also fulfill this function. It produces a kind of energy with low entropy that can be transported over a great distance, and it can also be converted into a variety of other forms of energy, such as motion energy, heat energy, and so on.

Electric power can be divided into alternating current (AC) power and direct current (DC) power. The characteristics of the current are what determine how the electric power should be categorized. Electricity is often priced per joule, which is calculated by multiplying the amount of power, measured in kilowatts, by the amount of time, in hours, that the equipment is operational.

The value of power can be determined by looking at an electric meter, which keeps track of the entire amount of energy employed by the devices being powered. The equation found below can be used to determine the amount of electric power.

Electrical Power= Work done in an electrical current/ time

$$P = \frac{Vlt}{t} = VI = IR^2 = \frac{V^2}{R}$$

In this equation, V represents the voltage in volts, I represents the current in amperes, R represents the resistance given by the powered devices, t represents the time in seconds, and P represents the power in watts.

Units

Watts are the measure of power used in electrical systems.

If,

V= 1 volts and I = 1 ampere

P= 1 watt

When there is a potential difference of one volt applied across an electrical circuit, the power used in the circuit is stated to be one watt if there is a current flow of one ampere in the circuit simultaneously. Kilowatts (kW), the larger unit of electrical power, are typically utilized in power systems because of their widespread application.

3.4 Types of Electric Power

The primary division of the available electrical power is into two categories. There is direct current power and alternating current power.

DC Power

The direct current (DC) power is described as the result of the direct current (DC) voltage. The generator, the battery, and the fuel cell all contribute to its production.

P= V x I

Where P – power in watts.

V – voltage in volts.

I – current in ampere

AC Power

The primary distinctions between the three varieties of AC power are as follows: The apparent power, the active power, and the true power are the three types.

Apparent Power

The power being used but not doing anything useful is referred to as apparent power. It is denoted by the sign S, and the volt-ampere is their SI unit of measurement.

$$S = V_{rms}I_{rms}$$

Where S – apparent power V_{rms} – RMS voltage = $V_{peak}\sqrt{2}$ in volt. I_{rms} – RMS current = $I_{peak}\sqrt{2}$ in the ampere.

Active Power

The active power, denoted by the letter "P," is the actual power lost due to the resistance of the circuit.

$$P = V_{max}I_{max}Cos\emptyset$$

Where P is the real power in watts

V_{rms} is RMS voltage = $V_{peak}\sqrt{2}$ in volts.

I_{rms} is RMS current = $I_{peak}\sqrt{2}$ in the amp.

φ is the impedance phase angle between voltage and current.

Reactive Power

It is the power derived from the reactance of the circuit. This type of power is called "reactive power," and it is created when the circuit reacts (Q). Volt-ampere reactive is the unit of measurement for it.

$$Q = V_{rms}I_{rms}Sin\emptyset$$

Where Q is the reactive power in watts

V_{rms} is RMS voltage = $V_{peak}\sqrt{2}$ in volts

I_{rms} is RMS current = $I_{peak}\sqrt{2}$ in the amp

φ is the impedance phase angle between voltage and current.

The formula below shows the relationship between the apparent, active, and reactive power parameters:

$$S^2 = Q^2 + P^2$$

The term "power factor" refers to the ratio of real power to perceived power, and the value of this ratio can range anywhere from 0 to 1.

3.5 Thevenin's Theorem

According to Thevenin's theorem, a single source voltage and a series of resistances linked across the load terminal can replace any linear circuit of a variety of resistances and voltages to create an equivalent circuit. This resistance is connected in series with the source voltage.

For instance, we may easily discover the balancing condition in intricate bridge circuits by reducing the circuit to its Thevenin equivalent. It is possible even in extremely complicated bridge circuits.

We don't have to identify the power dissipated in each element to calculate the average power lost in the circuit; we can do that before we figure out how much current is flowing through each element.

It is then possible to transform any electrical circuit and make it a two-terminal equivalent circuit with a single constant voltage source in series with a resistor connected to a load. It can be done by converting the circuit to an equivalent two-terminal circuit.

Advantages

The Theorem of Thevenin has some benefits:

1. It makes the less important element of the circuit easier to understand and allows us to see the behavior of the output component more straightforwardly.

2. It simplifies the circuit by reducing it to its most fundamental form, which is a single source of emf connected in series with a single resistance.

3. The theorem is extremely helpful for determining the amount of current flowing through a certain branch of the network when that branch's resistance is variable. On the other hand, every other resistance and emf supply is maintained at the same level.

The procedure of Thevenin's Theorem

If the network/circuit has several sources and resistances, finding the response of one element will take longer than usual. However, we may apply Thevenin's theorem to quickly get the answer. Let's use this theorem to find an element's reaction when there are several sources and resistances in the network/circuit.

1: Remove the element from the specified circuit where we are meant to find the response. The terminals will be open after the piece is removed.

2: Determine the voltage across the circuit's open-circuited terminals from Step 1. This value is called open-circuit voltage, or Vth.

3: In the circuit created in Step 1, replace all of the independent sources with their internal resistances.

4: If there are no dependent sources, find the resistance comparable between the open terminals of the circuit acquired in step 3 indirectly. This equivalent resistance is referred to as Thevenin's equivalent or Thevenin's resistance, abbreviated RTh.

5: If dependent sources are present, we can use the test source approach to calculate the equivalent resistance between the open terminals of the circuit determined in Step 3. We will connect a 1V (or 1A) source across the open terminals and determine another parameter current (or voltage) in the test source approach. The ratio of the current to the voltage between the two terminals will give us the value of Thevenin's resistance, RTh.

6: Create Thevenin's equivalent circuit by connecting Thevenin's voltage, VTh, and resistance, RTh.

7: Connect the element at the open terminals of Thevenin's equivalent circuit generated in Step 6 to the open terminals of the element where we expect to discover the answer.

8: Determine the element's response using simple laws or guidelines. Here, $(\text{Vth}/R_{th} + R)$ & $V= V_{th} (R/R_{th} + R)$.

3.6 Norton Theorem

To solve this particular electrical system, we use a theorem known as Norton's Theorem, which is also sometimes referred to as the Circuit Theorem, which is widely considered to be among the most fundamental network theorems. The depiction of a particular electric circuit as its equivalent circuit can be made more straightforward with the assistance of this handy theorem.

This part presents Norton's theorem and explains how to create Norton's equivalent circuit for a given circuit using the information presented in the previous section. The Norton theorem will be our first topic of discussion. After that, we will discuss the steps involved in formulating and applying Norton's theorem.

Theorem proposed by Norton: Any two-terminal linear and bidirectional network or circuit that has various independent and dependent elements can be expressed in a simpler equivalent circuit referred to as Norton's equivalent circuit, according to the expression that is part of Norton's theorem.

Norton's corresponding circuit is made up of Norton's power supply, I_N, and Norton's resistance, R_N. A realistic current source is a parallel combination of a current source and a resistor. As a result, Norton's analogous circuit is nothing more than an actual current source.

The Procedure of Norton Theorem

If the network or circuit has several sources and resistances, then the process of determining the response of an element will take significantly more time than would be required using the typical approaches. At that point, we can obtain the solution quickly by applying Norton's theorem. Now, let's have a look at the steps that need to be taken to find the response of an element while applying Norton's theorem and having many sources and resistances present in the network or circuit.

1: Remove the component where the answer from the specified circuit is meant to be found. The connections will be open after the piece is removed.

2: Determine the current flowing through the connections of the circuit generated in Step 1 after they have been shorted. This current is referred to as short circuit current, Norton's equivalent current, or Norton's current, I_N.

3: In the circuit created in Step 1, replace all of the independent sources with their internal resistances.

4: If there are no dependent sources, measure the equivalent resistance across the open-circuited endpoints of the circuit acquired in Step 3 indirect means. This equivalent resistance is referred to as Norton's equivalent resistance or Norton's resistance, abbreviated as R_N.

5: If dependent supplies are present, we can use the Test source approach to calculate the equivalent resistance throughout the open-circuited endpoints of the circuit acquired in Step 3. The test source approach involves connecting a 1V (or 1A) source across the open terminals and calculating another parameter current (or voltage). We shall calculate Norton's resistance, R_N, by dividing the voltage and current across the two terminals.

6: Create Norton's comparable circuit by attaching Norton's current, I_N, and resistance, R_N in parallel.

7: Connect the element containing the response to the open connections of Norton's equivalent circuit established in Step 6.

8: Employing laws or basic rules, determine the behavior of that element.

Formula

Norton's Theorem would be as follows for the above-mentioned circuit:

$$I = I_N(R_N/R+R_N) \ \& \ V = I_N \ (RR_N/R+R_N \ ).$$

3.7 Millman's Theorem

An electrical network theory is a useful technique for resolving complex electrical problems. It was named after its creator, electrical engineering professor Jacob Millman. Millman's Theorem, based on Thevenin's and Norton's theorems, is beneficial in simplifying complicated electrical circuits.

Millman's theorem, known colloquially as the Parallel Generator Principle, is useful for estimating voltage across a load and current flowing through a load. It can be applied in two ways: to circuits with only voltage sources or to circuits with both voltage and current generators.

Using this method, we can identify the voltage present across the network's parallel branches. This strategy simplifies the network even further when multiple sources are connected to it at the same time.

A circuit with voltage sources connected in parallel and an internal resistance can replace a circuit with a comparable voltage source connected in series with a resistance value.

How to Apply Millman's Theorem

We may provide a concise summary of how the theory should be applied to a real-world electrical network by following a few straightforward steps. The following are the steps:

1: In the first step of the process, the circuit transforms all voltage sources into current by dividing their potential by their internal resistance, where G is the conductance of the circuit.

$I=E/R=EG$

$G=1/R$

2: The second step is to use the following formula to determine the total current in the circuit.

$I_{eq}=E_1/R_1+E_2/R_2+E_3/R_3+....+E_N/R_N$

$I_{eq}=E_1/R_1+E_2/R_2+E_3/R_3+....+E_N/R_N$

3: In the Third step, mathematical equations can be used to determine the conductance of each current source.

$1/R_{eq}=1/R_1+1/R_2+1/R_3+....+1/R_N$

$G_{eq}=G_1+G_2+G_3+....+G_N$

4: To proceed to Step 4, we use Ohm's law to determine the voltage at the free end of the parallel circuit.

5: Fifth, determine the amount of current RL can handle.

$I_L=E_{eq}/(R_{eq}+R_L)$

Millman's Theorem Benefits and Uses

A complicated network can be broken down using Millman's theorem. Here are some frequent uses and benefits of the theorem.

1. When applied to a circuit with both voltage and current sources, Millman's theorem transforms voltage sources into current sources and current sources into voltage sources.

2. If there are enough voltage sources to prohibit a solution based on the series-parallel reduction approach from working, then Millman's theorem can be used to calculate the voltage across a collection of parallel branches.

3. Because it does not rely on instantaneous equations, Millman's theorem can be easily used by anyone. Complex topology circuits frequently employ numerous Op-amps for their various functions.

Problems With Millman's Theorem

Millman's theorem is quite general. However, it does not apply to all circuits without additional conditions. For this theorem to be useful, the following prerequisites must be met:

1. Impedances between unrelated components are outside the scope of Millman's Theorem.
2. In a circuit with independent and dependent sources, Millman's Theorem does not hold.
3. When a circuit contains two unrelated sources, this theorem is of no use.

3.8 Superposition Theorem

According to the superposition theorem, the output across any member in a linear, bidirectional network in which more than one source is present is equal to the sum of the feedback obtained from each source when they are analyzed individually. It holds in every network. On the other hand, their internal resistance takes the place of all other potential sources.

Next, we tell more about the superposition theorem, including cases that have been solved and the restrictions it imposes.

The superposition theory is a circuit simulation theorem used to solve networks that contain two or more sources coupled to each other.

The following is an example of what the superposition theorem says:

"The response of a component will be proportional to the algebraic sum of the response of that element when examining one supplier at the moment in any straight and bidirectional network or circuit that has many independent sources".

To determine the specific contribution made by each source in a circuit, it is necessary to eliminate or replace one of the sources in the circuit while maintaining the same overall result. To achieve this, the voltage source is first removed and then replaced with a short circuit. When a voltage source is disconnected, the value of the voltage is reset to zero. When a current source is removed, the value of the source is changed to infinite. To accomplish this, the current source is switched out for an open circuit.

Because it simplifies complex circuits by transforming them into their Norton or Thevenin equivalents, the superposition theorem is an extremely useful tool in the field of circuit analysis.

When applying the superposition theorem, keeping the following guidelines in mind is important.

1. Be very careful when allocating indications to the amounts as you add up the individual contributions that come from each source before adding them all up. It is recommended that a reference direction be assigned to each of the unknown quantities. An input from a resource is assigned a positive sign in the total if its direction is the same as the reference direction; alternatively, it is assigned a negative sign if its direction is the opposite of the reference direction.

2. To apply the superposition theorem to circuit currents and voltages, all of the components must be linear.

3. Because power is not a linear number, the superposition theorem does not apply to power, and this fact should be brought to your attention.

How Should We Put the Superposition Theorem Into Practice?

1. The first task is to choose one of the many different sources that are part of the bilateral network. Any one of the many sources in the circuit could be taken into consideration first. It doesn't matter which one it is.

2. Every source, except for the chosen one, has to have its external impedance substituted by its internal resistance.

3. Either the current flowing through a particular aspect in the network or the voltage drop occurring across that element should be evaluated using a network minimization technique.

4. When examining each of the other sources in the circuit, you must apply the same logic as when analyzing a single source.

5. After you have determined the appropriate response for each source, you must next undertake the operation of adding up all of the responses to get the total voltage drop or current flowing through the circuit element.

Limitations of Superposition Theorem

1. Non-linear circuits are excluded from the scope of this theorem. Because linearity is a prerequisite for using the superposition theorem to estimate anything, including voltage and current, it cannot be employed to calculate power. When only one source is taken into consideration at a time, power dissipation does not add up to an exact total since it is a nonlinear function and algebraic addition does not work on it.

2. To successfully apply the superposition theorem, the circuit must contain at least two different sources.

3.9 Nodal Analysis

In electrical networks, one type of method called "nodal analysis" is used to study circuits by looking at the voltages at the nodes, which are analogous to the variables in the circuit. This method is also known by its other name, the Node-Voltage Method. The following is a list of the primary characteristics of a nodal analysis:

Kirchhoff's Current Law (also known as the KCL) is an essential component of the Nodal Analysis.

If the number of nodes in the circuit is "n," then there will be "n-1" simultaneous equations that need to be solved.

It is possible to determine the voltage at each node in the circuit by working through the "n-1" equations.

The total nonreference nodes that can be produced are equivalent to the total nodal equations that can be found.

Both reference and nonreference nodes might be utilized in a nodal analysis project depending on the researcher's needs.

The Procedure of Nodal Analysis

When attempting to solve any electrical circuit by use of nodal analysis, the following procedures are to be followed in order:

1) Determine which of the primary nodes will serve as the reference node and then identify those principal nodes. This reference node will function as the ground for the network.

2) Label each of the other primary nodes' node voltages, except for the voltage at the reference node.

3) Identify the nodal equation for every principal node, except for the reference node. First, Kirchhoff's current law is used, and then Ohm's law is employed to complete the development of the nodal equation.

4) By following Step 3, we can derive the nodal equations, which will allow one to acquire the node voltages.

Therefore, the current flowing through any element and the voltage being applied across any element may be calculated using the node voltages for a specific electrical circuit.

Node Types in a Nodal Analysis

In nodal analysis, there are two different kinds of nodes:

1. A node without a reference
2. Node of reference

Node Without Reference

A non-reference node has a known node voltage.

Node of Reference

The reference node, sometimes called the datum node, is the reference node for all others.

There are two different categories of reference nodes:

- **Chassis Ground**

The chassis ground is a reference node for multiple circuits.

- **Earth Ground**

It is referred to as the earth's ground in any circuit where the earth's potential is employed as a reference.

Super Node

A voltage source attached between the two non-reference nodes, forming a generalized node, is called the super node.

Specifications of Super Nodes

Super nodes have the following characteristics:

1. In the super node, it is easy to calculate the voltage difference between two non-reference nodes.

2. Super nodes lack their internal voltage.

3. Kirchhoff's voltage law and Kirchhoff's current law are used to solve the super node.

3.10 Mesh Analysis

Even the simplest circuits have their importance in the field of electronics; principles such as KVL and KCL are used for their analysis. However, when dealing with complex circuits with several sources of regulated voltage and current, more techniques are required beyond KVL and KCL laws. Analysis based solely on KVL and KCL concepts is shown to be flawed and unreliable. Approaches such as mesh and nodal must be applied to go with detailed analysis and to understand the parameters in those circuits. Constants like current and voltage can be accurately tested using these techniques.

A loop that is deemed to be mesh is one contained within a circuit that does not have any other loops. To carry out an exhaustive study of the circuit, mesh currents have been substituted for currents as the variables that have been employed here. As a consequence of this, the method requires a condensed form of the equations to be solved. Kirchhoff's voltage law is utilized in the implementation of mesh analysis in the circuits to determine the unknown current values.

A different name for this approach is the "mesh current loop technique." Following this, the use of Ohm's law will allow for the determination of the values of the voltage. When a path joins two nodes in a circuit, the way is known as a branch, and the branch itself is regarded as an element of the circuit. When there is only one branch connecting two nodes in a mesh, we refer to the current flowing through that branch as the mesh current. On the other hand, if a mesh has two branches, the current through the mesh can be calculated as either the total or the difference of the currents flowing through the two mesh loops, depending on whether the two loops are traveling in parallel or opposing directions.

Steps for Using Meshing Analysis

To put mesh analysis into practice, we need to follow a process, and the steps involved are described below:

1. The initial step is to identify the meshes and then label the mesh currents in either the counterclockwise or the clockwise direction, depending on which is appropriate.

2. Investigate the total amount of current that moves through each element as it relates to the mesh currents.

3. Take note of all of the mesh equations for the meshes that were seen. The equations for the mesh are written by first applying Kirchhoff's law and then Ohm's law.

4. To determine the mesh currents, one must first solve the measured mesh equations following the step.

5. By applying mesh currents, it will be possible to determine the flow of current and the voltage readings across every component in the circuit. This is made possible thanks to the previous point.

General Method for Creating Equations in Mesh Analysis

Every single mesh in the circuit can be broken down into a single equation once it has been identified. These equations represent the sum of the voltage drop that occurs along the entirety of the mesh current's loop. In the circumstances of circuits that have more than just voltage and current, the voltage drop is regarded as the circuit's impedance, which is combined with the specific loop mesh current. It is done to calculate the resistance of the circuit.

When the voltage source is located within the loop itself, the voltage already existing at the source can either be added to or removed from the total depending on whether the situation being addressed is a decrease or an increase in voltage for that mesh. In contrast, if the source of the current does not lie in between the meshes, the direction of the mesh current will determine whether a negative or positive value is assigned to the source by the mesh current.

Super Mesh Analysis

The analysis of enormous and complicated circuits is best accomplished through the use of super mesh analysis, as opposed to traditional mesh analysis. It happens because super meshes consist of two meshes working together to share a component that serves as the current source.

Because this method simplifies those complicated circuits by closing the voltage element and reducing the number of reference nodes for each voltage source, it is used as an alternative to the technique that is followed for nodal circuit analysis. This technique is known as super node circuit analysis. Within the super mesh analysis context, the current source is located within the super mesh section. This placement allows us to reduce the number of meshes by one for every current source present.

If the source of the current is located on the permitter of the circuit, then a single mesh should not be considered. On the other hand, the KVL algorithm is only applied to the meshes included in the updated electric circuit.

To gain a better understanding, let's look at a real-world example of super mesh analysis.

Uses of Mesh Analysis

The most important application of mesh analysis is to resolve planar circuits to determine the current values at every place in electric circuits, no matter how simple or complex they may be.

Another application is that traditional computations to solve equations are difficult and require more mathematical formulas, but by using mesh analysis, we only need to perform a sufficient number of computations.

The unbalanced wheat stone bridge is another application for the Mesh Current analysis technique. Take a look at the sample below to see how this works.

Because the ratios of the resistors R1 divided with R4 and R2 divided with R5 are not the same, we can understand that there will be a certain amount of voltage as well as current flow at R3. Since we are aware that solving problems involving this type of circuit is made more difficult by using the general series-parallel technique, we need a different approach to handle this problem.

In light of this, one option available is to use the branch current approach; however, this method requires six currents to flow from Ia to If, which means that it can be applied to any number of equations. Because of this, the complexity of the problem can be easily decreased by using the present method of meshing, which only requires a few variables.

3.11 Kirchhoff's Current Law

KCL, states that "the total of all currents that arrive and depart should sum to zero." KCL is an abbreviation for "Kirchhoff's Current Law." This statute is usually shortened to its abbreviated form.

This law characterizes how a charge enters and exits a junction point or node on a wire. A node is a location where two wires join.

Currents In and Out of a Node

Through each node along the positive "rail," there is a current break that occurs from the main flow to each successive branch resistor. The current from each succeeding branch resistor is pooled at each node along the negative "rail," which generates the main flow. This occurs as a result of the fact that all of the nodes are linked to the same wire. If you think of the hydrant circuit as an analogy, each branch node functions as a "tee" fitting, and the water flow either splits or merges with the main piping system as it travels from the outcome of the water pump toward the exchange reservoir. This reality should be quite obvious to you if you use the water pipe circuit as a resemblance.

To use the analogy of plumbing: Provided that there are no leaks in the pipes, whatever flow enters, the fitting must also escape the fitting. This is true as long as there are no leaks in the pipe. This is the case for any node (or "fitting"), regardless of the number of flows that are either entering or leaving. To explain this general relationship using mathematics, we can write it as follows:

$$1_{exiting} = 1_{entering}$$

Mr. Kirchhoff came up with the idea to rephrase it in a slightly different way:

$$1_{entering} + (-1_{iting}) = 0$$

Kirchhoff's Current Law can be summed up in one phrase: "The sum of all those entering currents and departing currents must equal zero."

That is to say, if we give each current a mathematical notation, which indicates whether they go into (+) or come out of (–) a node, then we can add up all of those signs and arrive at a sum that is certain to be zero.

We can calculate the intensity of the current that flows through the left side of the network by establishing a KCL formula using that current as the variable and using our instance node (number 6) as a starting place in the calculation:

$I_2 + I_3 + 1 = 0$

2ma+3ma+1=0

For solving I:

1=−2mA − 3mA

I=−5mA

The current is leaving the node when we look at the value of 5 milliamps because it has a negative sign (−) next to it. It contrasts with the 2-milliamp and 3-milliamp currents, which should be positive. It makes no difference whether a negative or positive sign represents current entering or departing; what matters is that they are opposite signs for different directions and that we remain consistent in our notation. If we do those two things, KCL will function correctly.

Kirchhoff's Voltage and Current Laws are a powerful combination of instruments that can be utilized in examining electric circuits when used together. The electronics student should commit these laws to memory just as much as they should commit Ohm's Law to memory.

3.12 Ohm's Law

Ohm's law indicates the relationship between current, resistance, and voltage; it is the most often utilized principle in electrical systems.

VOLTAGE = CURRENT * RESISTANCE

Voltage and current have a linear connection, which means that when voltage increases, so does current, and if voltage drops, then current also drops. The current that will travel through a circuitry with a particular load can be easily calculated.

Ohms' Law in Other Forms:

Current=Voltage/Resistance

Resistance=Voltage/Current

Chapter 4:
The Importance of Testers

4.1 Different Testers and Indicators

The simplest electrical testing method involves providing a voltage or current to a circuit and comparing the observed value to the anticipated outcome. Each piece of electrical test equipment is designed for a particular use and checks the mathematics behind a circuit.

A test technician's responsibility is to understand the constraints of the test equipment they are using and know which test instrument to employ for the work. We examine the most prevalent test instruments in use today in this chapter.

4.2. Megohmmeter

A megohmmeter, often known as a "megger," is a unique kind of ohmmeter used to measure insulators' electrical resistance.

Megohmmeter resistance readings may vary from a few megohms to several million megohms (teraohms). Megohmmeters generate high voltages manually or by battery-powered internal circuitry with outputs ranging from 250 to 15,000 volts.

Megohmmeters, among the most regularly used test instruments, may be employed to gauge the insulation of various equipment, including cables, transformers, switchgear, and circuit breakers.

4.3. Multimeter

A multi-meter, sometimes called a VOM (Volt-Ohm meter), is a portable instrument that integrates various measuring functions into a single unit.

Multi-meters are primarily used to diagnose electrical issues in various commercial and home appliances, electronic equipment, motor controllers, power supply, and wiring systems.

Although analog multi-meters are still preferred in certain circumstances, such as when tracking a fast-changing value or delicate measures, such as checking for CT polarity, digital multi-meters are still the most often used.

4.4. Low-resistance Ohmmeter

The low-resistance ohmmeter, often known as a DLRO in the field, is used for high-precision resistance measurements below 1 ohm. Low-Resistance ohmmeters have battery power outputs of up to 100A and can generate low-voltage DC currents.

With the use of four terminals, referred to as Kelvin contacts, resistance measurements may be made. Two terminals (C1, C2) enable the meter to measure the voltage across the resistor, while the other two carry the current from the meter (P1, P2). With this kind of meter, the meter ignores any voltage drop caused by the resistance of the first pair of leads.

One of the most used pieces of test equipment, low-resistance ohmmeters, may be used to measure the resistance of various devices, including fuses, transformers, motor windings, cable and busways, and switch and circuit breaker contacts.

4.5. High Current Testers

The "control unit" and "output unit" of a high current test set may be two separate components, or these features may be included in a single package. Low voltage circuit breakers are tested using primary injection with low voltage, high current outputs.

Large transformers used in the high current or "primary-injection" test set down line voltage (480V) to a very low level, such as 2–15V. A significant increase in the available current output (15 kA+), particularly for a short period, is made possible by the significant voltage drop.

A variable resistor and tap changer regulate the current output. A circuit breaker's trip time is shown by integrated timers as the interval between current on and current off.

Circuit breakers may be directly connected by bus or cable to the high-current test set. By connecting directly to the switchgear bus, this test equipment may also be used to test ground faults and other current relays, depending on its size.

4.6. HI potential Testers

In contrast to a continuity test, hi-pot testing examines the integrity of the insulation in medium- and high-voltage equipment. To ensure there are few current leaks from the insulation to the ground, insulation is strained above nominal levels.

A high-voltage lead, a ground lead, and a return lead make hi-pot test sets. The tested device is connected to the high-voltage lead, and all other parts are grounded. The resultant current is then measured via the return.

The test set's inbuilt safeguard will trigger if there is an excessive flow of return current. Leakage current must not trigger the test set for the hi-pot test to pass; nevertheless, there is no minimum permissible value.

Depending on the equipment being tested, the output voltage might vary from 1kV to 100kV+ at line frequency or dc. An AC sinusoidal waveform, typically at 0.01 to 0.1 Hz, is used in very low frequency (VLF) to withstand testing to evaluate the effectiveness of electrical insulation in heavy capacitive loads, such as cables.

4.7. Relay Testers

These power system simulators test protective mechanisms for power and industrial systems. Relay test sets are equipped with several sources to evaluate solid-state and multi-function numerical protection. Each voltage and the current channel are controlled independently to provide various power system circumstances.

In addition to testing complicated protection systems like communication-assisted line protection and protection schemes that utilize IEC61850-compliant IEDs, high-end relay test equipment can also test straightforward voltage, current, and frequency relays (intelligent electronic devices).

4.8. Power Factor Testers

With the help of Power Factor Examination Sets, high voltage equipment, including transformers, bearings, circuit breakers, cables, spark arrestors, and spinning machinery, may undergo a thorough AC insulation diagnostic test.

The power factor test set monitors the voltage and current of the device under test using a standard impedance, and test voltages are typically 12 kV or less. The vector voltage and current are used to generate every result that has been published, including those for power loss, power factor, and capacitance.

A specimen's capacitance and dissipation factor (power factor) are measured during tests. When unfavorable circumstances occur, such as moisture on or within the insulation, conductive impurities in insulating oil, gas, or solids, internal partial discharges, etc., the values measured will vary.

A single high-voltage lead, two low-voltage leads, and a ground are included in test connections. For the protection of the operator, safety switches and strobe lights are provided, and test readings are corrected using a temperature sensor. An Ethernet or USB-connected laptop is often used to run power factor test setups.

4.9. *Winding Resistance Testers*

Winding resistance measurements are a crucial diagnostic tool for evaluating potential damage to the transformer and motor windings. Shorted turns, weak connections, or worn-out tap changer contacts will cause modifications in the winding resistance of transformers.

The voltage drop at each terminal is measured after a known DC has been sent through the winding being tested (according to Ohm's Law).

A winding resistance test set is similar to a large, low-resistance ohmmeter; modern test equipment for these applications uses a Kelvin bridge to provide findings (DLRO).

Test sets for winding resistance feature two current leads, two voltage leads, and one ground lead. A test set for winding resistance typically has a current range of 1A to 50A. On secondary windings with high currents, it has been discovered that higher currents shorten test periods.

4.10. *Secondary Testers*

It is possible to test circuit breakers with solid state and microprocessor trip units by directly injecting secondary current into the trip unit rather than utilizing a high current test set to transfer primary current through the CTs. The secondary current injection test method's primary drawback is that it only tests the solid-state trip unit logic and components.

Manufacturers of trip units provide secondary test sets specifically for use with a single type or family of trip units using a proprietary connection. Test kits may vary in sophistication from straightforward "suitcase" devices that function similarly to a principal injection test set to basic hand-held, push-button designs.

Handheld devices often circumvent trip unit protection features like ground fault when testing circuit breakers using primary injection.

4.11. Transformer Turns Ratio Testers

The turns ratio (TTR) test set applies electricity to a transformer's high-voltage winding and measures the voltage that emerges from the low-voltage winding. The devices also monitor excitation current, and percent ratio error in addition to turn ratio.

Transformer turns ratio test cases are available in many designs and test connections, but they always feature a minimum of two high and two low leads. A TTR test set's excitation voltage is typically less than 100V.

4.12. Current Transformer Testers

CT test sets are compact, multi-purpose devices used to test current transformers for demagnetization, saturation, ratio, winding resistance, polarity, phase deviation, and insulation. High-end CT test tools can easily connect to multi-ratio CTs and instantly run all tests on all taps without switching leads.

Current transformers may be evaluated in equipment arrangement, such as when placed in transformers, oil circuit breakers, or switchgear. A contemporary CT with multiple voltage and current outputs may be a relay test set when used with a laptop computer.

4.13. Ground Resistance Testers

The ground resistance test set operates by sending a current into the ground via a test electrode and a remote probe, measuring the voltage drop brought on by the soil at a predetermined location, and then calculating the resistance using Ohm's Law.

The most popular ground resistance test sets are the 4-terminal unit for evaluating soil resistivity and the 3-terminal unit for measuring fall-of-potential. Spools of tiny stranded wire are used to cover long-distance measures with copper rods or stakes of a similar design to establish contact with the soil.

They come with limits but provide precise data without cutting off the ground system being tested.

4.14. Power Recorder

Power recorders are tools for gathering voltage and current data that may be downloaded into software to examine the state of an electrical system. These diagnostic techniques identify electrical issues such as voltage swells, sags, flickers, and low power factors.

Engineers who want to develop a system or consumers who want to check their energy bills may benefit from using power recorders to quantify power use over time. Power recorders come in various sizes, degrees of precision, and storage capacities.

Wrapping conductors with split-core CTs and connecting a set of leads to system voltage and ground are required to install a 3-phase power recorder. A PC or built-in screen may be used to see the recorder in real-time when it is configured to measure by the system configuration for a certain amount of time.

4.15. Infrared Camera

The infrared radiation—undetectable to the naked eye—is detected by thermal imagers, turning that information into a colorful image on a screen. Because test processes are non-contact and may be completed rapidly with equipment in operation, infrared cameras are most often employed to examine the integrity of electrical systems.

An effective method of troubleshooting is to compare the thermal signature of a piece of equipment in regular operation to the equipment being checked for abnormal conditions. An aberrant thermal picture might be utilized to evaluate whether more testing might be necessary, even if it cannot be completely comprehended.

The accuracy and sensor resolution of thermal imagers is used to categorize them. High-end infrared cameras can collect high-resolution images and have temperature precision of at least one-tenth of a degree.

4.16. MAC Testers

Traditional vacuum interrupter field testing uses the hi-potential test to assess the bottle's dielectric strength; however, this test only yields a go/no-go result, not indicating if or when the gas pressure within the bottle has fallen to a critical level. In contrast to the hi-pot test, testing vacuum interrupters based on magnetron atmospheric condition (MAC) principles may provide a practical way to identify a vacuum interrupter's condition before failure.

The vacuum interrupter is only inserted into a field coil to set up the magnetic field test, which will result in a DC that is consistent during the test. The open contacts are subjected to a continuous DC voltage, typically 10 kV, and the current flowing through the VI is monitored.

4.17. Vibration Analyzer

The most frequent mechanical problems (bearings, misalignment, imbalance, and looseness) in rotating equipment are identified and located using vibration analyzers.

Vibration levels rise when motors acquire mechanical or electrical defects. These increases in noise and vibration occur at various stages of a growing fault.

With the machinery in operation, acceleration measurements are made using accelerometers, and the data is then put into the analysis software. The accelerometer measures the device's vibration being tested in three planes of motion (vertical, horizontal, and axial).

4.18. Ultrasound Analyzer

All three arcing, tracking, and corona phenomena result in ionization, which affects the nearby air molecules. These emissions produce high-frequency noises, which an ultrasonic tester picks up and converts to human hearing ranges.

Each sound emission may be heard with headphones, and a display screen shows the signal's strength. These sounds may be captured and examined using software for ultrasound spectrum analysis for a more precise diagnosis.

Electrical devices should typically be quiet, yet some, like transformers, may emit a continuous hum or other mechanical sounds. These should not be mistaken for an electrical discharge with unpredictable, uneven, popping, sizzling, frying sounds.

Additionally, air leaks in transformer tanks and gas-insulated circuit breakers may be found using ultrasonic detectors.

4.19. Testers for Battery Impedance

Cell impedance, voltage, inter-cell connection resistance, and ripple current are some of the key battery characteristics that may be measured using battery impedance test equipment, which is mostly employed in substation and UPS applications to assess the health of lead-acid cells. A single unit can handle all three tests.

When using the battery impedance tester, each cell is crossed by an AC signal, and the AC voltage drop and the current in that cell are measured. The impedance will then be calculated. Dual-point, Kelvin-style lead sets are the norm. One point is used to apply current, while the other is used to measure potential.

4.20. Load Bank

Electric power sources like diesel generators and uninterruptible electricity supply are commissioned, maintained, and verified using load banks (UPS). The load bank subjects the test equipment to an electrical load, releasing the electrical energy as heat via resistive parts. The load bank architecture uses motorized fans to cool the resistive parts.

If necessary, many load banks may be linked together. A load bank may be entirely resistive, entirely inductive, entirely capacitive, or any mix of the three. The greatest approach to duplicate, demonstrate, and verify the actual demands on essential power systems is through load banks.

4.21 How Insulation Resistance Testers Work

The earliest and most used test for measuring the quality of insulation is the insulation resistance (IR) test, which was developed in the early 20th century. The second test required for electrical safety is the insulation resistance test. The Insulation Resistance Test involves short-circuiting the phase and neutrality of a device while measuring its insulation resistance. The measured resistance must be higher than the cap set out by international standards.

Similar to how friction cannot exist, insulation cannot be flawless. This suggests that some power will be present there continually. This is known as leakage current. Leakage is OK with enough insulation, but issues may start to appear as the insulation deteriorates. What therefore makes insulation "good"? It must be capable of maintaining a significant length of time with a high current resistance.

The Purpose of an Insulation Resistance Test

As soon as the insulation is produced, it begins to deteriorate. It loses insulating effectiveness as it becomes older. This process is accelerated by severe installation settings, particularly those with temperature extremes and chemical pollution. Stresses are brought on by several circumstances, such as:

- Overvoltage and undervoltage are the main causes of electrical stressors.

- Mechanical strains may be brought on by repeated start-up and shut-down procedures.

- Issues with machinery's balance and any direct load on the wires and installations.

- Chemical stresses: The materials' ability to insulate is impacted by the presence of chemicals, oils, corrosive vapors, and dust.

- Stresses related to changes in temperature: Expansion and contraction stress impact the insulating materials' characteristics when paired with the mechanical stresses brought on by the start-up and shutdown procedures. The materials age due to operation in very hot or cold conditions.

Insulation ages more quickly because of environmental pollution.

This wear and tear may lower the electrical resistivity of the insulating materials, increasing leakage currents that might result in accidents that could be costly in terms of production stoppages and safety (for both people and property). Therefore, it's critical to spot this decline as soon as possible so that remedial action may be implemented. These tests identify aging and early degradation of the insulating qualities before they reach a point where they are likely to result in the occurrence of the accidents above.

This test is used as an acceptance test, and the client frequently specifies the insulation resistance unit length. The findings of the IR Test are meant to provide information on the quality of the bulk material used as the insulation rather than helping identify specific localized faults in the insulation as in a real HIPOT test.

Wire and cable manufacturers use the insulation resistance test to monitor the progress of their insulation production processes and identify emerging issues before variables depart outside the permitted range.

Insulation Resistance Measurement

All different kinds of electrical wires and cables are routinely tested by measuring insulating resistance. Its goal is to determine the insulation's ohmic value at a high-stability direct voltage, often 50, 100, 250, 500, or 1000 VDC. Megohms (M) represent the insulating resistance's ohmic value. The insulation resistance test may be carried out at voltages up to 1500VDC to meet specified requirements. The voltage source's reliability makes it feasible to change the test voltage in increments of 1 volt.

The voltage must be stable; otherwise, it will decrease suddenly due to poor insulation, leading to inaccurate measurements.

After making the necessary connections, you apply the test voltage for one minute. The resistance should decrease or stay mostly stable over this time. Smaller insulation systems maintain stability, whereas larger ones exhibit a continuous decline because their capacitive and absorption currents reach zero more quickly. Read and note the resistance value after one minute.

How to Measure Insulation Resistance

The insulation resistance is measured using an IR tester. This portable instrument features a built-in generator for high DC voltage and resembles an ohmmeter. Current flows across the surface of the insulation as a result of the voltage, normally at least 500V. It offers an ohm-based IR measuring.

The basis for calculating insulating resistance is Ohm's Law. (R=V/I). By applying a known DC voltage lower than the voltage for dielectric testing and then watching the current flow, it is comparatively simple to determine the value of the resistance. The insulating resistance value is theoretically infinite, but the megohmmeter monitors the low current flow to calculate it, delivering the result in kW, MW, GW, and TW (on some models). This resistance is a trustworthy gauge of the risks caused by leakage currents and reveals the efficacy of the insulation between the two conductors.

If you see a lot of IR, your insulation is probably quite good. On the other hand, if it is low, the insulation is insufficient.

However, this is not the only factor that could affect the IR; a few others are temperature and humidity. You will need to do several tests over time to ensure that the IR value stays substantially the same. The giga ohm is the unit of measurement for insulating resistance (G).

Good insulation is indicated by megger readings that increase initially before holding constant. Megger readings that increase at first and then decrease indicate insufficient insulation.

Between 20 and 30 degrees Celsius are the expected IR values. If this temperature dips by 10 degrees Celsius, IR measurements will double. If the temperature increases by 70 degrees Celsius, the IR measurements decrease by 700 times.

The measuring voltage must be substantially higher than typical resistance quantifications to estimate big electrical resistance. Electronic components cannot be harmed by this voltage, which is frequently between 100 VDC and 1000 VDC, so it cannot be used to measure their resistance.

Dielectric Strength Testing

The "breakdown test," also known as "dielectric strength testing," gauges the insulation's capacity to sustain a medium-duration voltage spike without spark over. This voltage spike may have been generated by lightning or an electrical transmission line malfunction that caused induction. This test's primary goal is to confirm that the building regulations regarding leakage routes and clearances have been adhered to. Although a DC voltage may also be used, an AC voltage is often employed to conduct this test. A hi-pot tester is necessary for this kind of measurement. The outcome is a voltage value often given in kilovolts (kV). Depending on the test levels and the instrument's available energy, dielectric testing may be harmful in the case of a malfunction. It is only used for type testing on new or refurbished equipment because of this.

Under typical test circumstances, however, insulation resistance measurement is non-destructive. It produces a result stated in kW, MW, GW, or TW after applying a voltage with a lower amplitude than for dielectric testing. This resistance indicates the caliber of the insulation between the two conductors. It is especially helpful for monitoring insulation aging throughout the operational life of electrical equipment or installations since it is non-destructive.

Insulation Resistance Measurement Safety Requirements

All test-related equipment has to be unplugged and separated.

To ensure complete safety for the person doing the test, the equipment should be discharged for as long as the test voltage has been exerted.

Never use Megger in a volatile environment.

Ensure all switches are closed off, and cable ends are appropriately marked for safety.

When checking for the earth, be sure the conductor's far end is not contacting; otherwise, the test will indicate poor insulation when none exists.

Verify the tightness of each connection in the test circuit.

The supply must be cut off from the cable ends that need to be isolated and shielded against accidental contact with the ground or the supply.

Construct safety barriers with warning signs and open lines of communication between test participants.

Megger's profile

Typically, a megohmmeter has three terminals.

The conductor is attached to the "LINE" (or "L") terminal, which is referred to as the "hot" terminal. Keep in mind that the circuit is not powered up throughout these tests.

The "EARTH" (or "E") connector is wired to the ground conductor on the insulation's opposite side.

Bypassing the meter, the "GUARD" (or "G" terminal offers a return circuit. For instance, if you wish to exclude a current from the circuit you are measuring, you link that portion of the circuit to the "GUARD" terminal. The challenge at hand is the easiest one.

Why Can't Insulation Resistance Be Measured With a Multimeter

Electrical resistance, which unit is reported in ohms, is one of the several magnitudes that a multimeter can measure. An internal battery that circulates a tiny current through the resistance being measured or, if that isn't possible, the conductor or winding, is responsible for the device's functioning, notably for measuring resistance. The measured value in ohm represents the conductor's electrical resistance, increasing following its longitude and section and allowing current to flow through it.

On the other side, a megohmmeter is frequently used to gauge an insulated body's insulation resistance. Depending on the application, it may create output voltage levels of up to 5000V using a battery or a DC generator. The ohm test yields findings linked to insulation resistance, with an insulated element connected to an active element or wire.

Despite some similarities between the two instruments, a Megger (or a similar device) is required to measure insulation resistance because it may produce a high voltage that causes a moment of stress in the insulation. In most cases, insulation resistance is estimated in mega- or tera-ohms, respectively.

In conclusion, a multimeter can only measure the insulating resistance of an isolated group, but a Megger can measure the electrical resistance of a conductor (coil).

Insulation resistance test types

- **Test of Spot or Short-Time Reading**

With this technique, you attach the Megger instrument to the insulation to be tested and run it for a brief, predetermined amount of time. You have merely chosen a spot on a curve of rising resistance values; often, the value would be lower for 30 seconds and higher for 60 seconds.

The spot reading test is all that is required if the instrument you are evaluating has very little capacitance, such as a short run of home wire. Maintenance specialists have established the acceptable lower limit for insulating resistance using the one-megohm guideline for many years. The rule may be expressed as follows: Insulation resistance should have a minimum value of one megohm and should be about one megohm for every 1,000 volts of operational voltage.

- **Method of Time Resistance**

This technique often provides conclusive information without previous test results and is mostly temperature-dependent. It is based on how well-insulated buildings absorb sound compared to those with damp or polluted insulation. This kind of examination is also known as an absorption test.

The fact that this test is unaffected by the equipment size adds value. Whether a motor is big or small, the rise in resistance for clean and dry insulation happens similarly. Therefore, regardless of their horsepower ratings, you may evaluate several motors and develop benchmarks for new ones.

Insulation Resistance should be performed to avoid risks like an electric shock and short circuits brought on by the insulation deteriorating with time in electrical components, equipment, and devices used in industrial facilities, buildings, and other settings.

4.22 Electrical Testers

1. Tester for Non-Contact Voltage (Inductance Tester)

Inductance testers allow you to check for voltage in cables and other devices without contacting the components themselves. They're not only cheap and easy to employ but also risk-free.

The device, which looks like a miniature wand with a pointed end, can be used to check for voltage in various electrical components, such as wall outlets, circuit breakers, and switches. The tester's tip can be inserted into the outlet slot or touched outside a wire or electrical cable to obtain a reading.

A red light at the end of most voltage testers and a buzzing sound indicate the device's voltage. The simplest forms reveal whether or not electricity is present. More advanced (and pricy) variants give a rudimentary evaluation of the voltage present, albeit the estimate is less reliable than that of some other electrical testers.

2. Voltage Tester for Neon

The voltage in a circuit cannot be determined by neon voltage testers; they can only determine if the voltage is there. They have two short wire lines with a metal probe on each end and a little body with a neon light inside. This gadget is trustworthy since it doesn't need a battery. It's also reasonably priced.

Place one tester probe against a hot wire, or screw terminal to utilize a neon voltage tester. Connect the second probe to a ground or neutral contact. The little neon bulb at the tool's tip will illuminate if there is a current.

The grounding of a plug can also be checked with a tester. If the outlet tests well when probes are inserted into the hot and neutral slots but turns dark when the neutral probe is transferred to the grounding location, the outlet is not properly grounded.

3. Circuit Analyzer Plug-In

When connected to a power source, plug-in circuit analyzers morph into low-cost, straightforward testers that provide a wealth of information about a circuit's inner workings. All three-prong grounded outlets can be checked with these handy gadgets. They won't fit into the standard two-slot outlets used on older appliances.

Three neon lights flicker in different sequences on plug-in circuit analyzers to represent test results. To identify the various light patterns, simply stick a chart to the tester. Each possible state of an outlet's wiring is represented by a different set of lights: properly wired, reversed, open, and grounded.

Simply inserting circuit analyzers into a light outlet allows them to run without batteries. The analyzer needed electricity to function.

The voltage and state of a circuit are displayed on an LCD screen by modern plug-in circuit analyzers. This new kind does not require electricity or recharge.

4. Tester of Continuity

One end of a continuity tester has a sensor, and the other end can be connected to an alligator clip or another sensor. An electrical wire often has two points of contact, one at each end. When the tester's internal light starts to shine, the circuit is complete as well. If a circuit is overloaded, some appliances may emit an audible signal.

In contrast to voltage testing, continuity testers are utilized whenever a cable or device is unplugged or a circuit is turned off. They don't test for voltage but rather for continuity in the electrical circuit of a gadget or appliance. They are useful for checking if a fuse has blown or if a single-pole or three-way switch is functioning properly.

Always disconnect the power supply to the circuit or equipment before using a continuity tester on it. Turn it off and disconnect it from the wall. It's risky to test an electrical wire with a continuity tester.

5. Multimeter

Electrical testers with a wide range of testing capabilities include a multimeter. Resistance, voltage, capacitance, and frequency can all be measured precisely with most multimeters. As a result, they can supply almost all the information that other kinds of electrical testers can.

The multimeter features two long leads with probes at each end, a box-shaped body with a digital or analog readout, a dial for configuring the test, and two probes. The quality and accuracy of these devices may vary greatly, and you often pay more for higher quality. They are often more costly than entry-level testers but are still affordable.

6. Tester for Solenoid Voltage

For testing voltage and polarity, solenoid voltage testers are less complicated to operate than multimeters. Because it is durable and requires no batteries, professionals often choose this equipment over the multimeter. However, it is less precise than a multimeter in giving a numerical reading of the voltage level. However, it often costs less than a multimeter.

Models are offered in both analog and digital formats. Two wires with a probe extending from the tester's bottom. They click or vibrate to signal the presence of electricity; the louder the clicking or the more intense the vibration, the greater the voltage level. During testing, they often GFCI circuit breakers

7. Clamp Digital Meter

A digital clamp meter is more expensive than a multimeter and combines the functionality of a multimeter with a current sensor. Only homeowners need it if they are doing complex electrical repairs; it is a specialized tool.

The functions of a multimeter and a clamp meter vary only slightly. The clamping jaws of this tool can hold onto wire conductors, which is the most evident advantage.

4.23 Electrical Testing and Inspection

Our daily activities would not be the same without electrical equipment. We utilize them in our homes and workplaces to make work easier and run our businesses more efficiently. However, we cannot disregard the potential threat to our safety and property if anything goes wrong. The importance of electrical testing and inspection will be covered in this section.

1. It Promotes Product Design Advancements

Fundamentally, electrical testing establishes if a product is suitable for usage.

Manufacturers did, however, learn more about their goods over time. Manufacturers saw the flaws in their product's design as stepping stones to a more perfected version rather than as obstacles. Electrical testing and inspection paved the way for creative concepts that improve product design while putting safety first.

2. It Lessens the Possibility of Damage or Injury When in Use

When utilizing electrical equipment, you risk suffering from shock, electrocution, fire, energy discharge, or burns. Additionally, moving elements or sharp edges might hurt you or crush you. Making sure the product design won't injure or damage any person, pet, or property is one of the main goals of electrical testing and inspection.

3. It Assures That You Adhere to All Legal Standards

Consumers often believe that electrical equipment is secure as long as the voltage is low.

A low-voltage gadget might be just as hazardous as one with high voltage.

You may carry out your legal responsibility to ensure a product is safe for consumption before it enters the market by using electrical testing and inspection.

4. It Leads to Better Test Result Documentation

A test report should be created each time a technician performs an electrical safety test to record their results. They put the data in a technical file after the report is finished. According to regulations, every product must have a technical file containing all of the safety testing results completed when it was first put up for sale and throughout the ten years after a change in ownership. Since less time is spent doing tests, electrical testing and inspection assist specialists in finding methods to enhance the documentation process.

5. It Instills a Lot of Confidence

A facility is subjected to quality inspections by a professional organization or authority to ensure the tests are up to grade. The evaluations confirm that the institution has engaged competent staff members who are trained to carry out the exams carefully.

Chapter 5:
Electricity and Safety

I t is crucial to a circuit's proper operation that the amount of electricity flowing through it be restricted. This safe level is determined by the amount of current that the load, conductors, switches, and other components of the system can safely carry. However, there are situations when an abnormally high amount of current will flow over a certain electrical circuit (overcurrent).

5.1 Electrical Safety Wiring

When working with electricity, it is necessary to exercise extreme caution and adopt all necessary safety measures. The importance of safety cannot be overstated, and before anything else, certain guidelines need to be adhered to.

The first step in maintaining your safety is to steer clear of any sources of water. Never attempt to repair any electrical equipment or circuits while your hands are wet; this includes touching the equipment itself. It makes the electric current more conductive. Never use any piece of machinery if the cables are frayed, the insulation is compromised, or the plugs are broken. If you are working on a receptacle in your home, you must always turn off the mains before beginning any job. A warning sign should also be affixed to the service panel to ensure that no one will inadvertently flick the main switch to the "ON" position. When working, you should always use insulated tools.

Electrical dangers include energized parts exposed to the environment and unsecured electrical equipment that has the potential to become energized without warning. Warning labels stating "Shock Risk" are consistently affixed to equipment of this type. Always keep an eye out for warning signals like these, and be sure to abide by the safety regulations outlined in the electrical code that is in effect in the nation you are in. When working on any branch circuit or any other electrical circuit, you must wear suitable insulated rubber gloves and eyewear at all times.

Never attempt to repair anything that is electrified. Always use a tester to ensure it has been de-energized before proceeding. When an electric tester touches a live or hot wire, the light-emitting diode (LED) contained within the tester illuminates to indicate the presence of an electrical current in the wire being tested. Before you continue with your job, use an electrical tester to examine all of the wires, the outside metallic covering of the service panel, and any other wires hanging.

If you are working on a receptacle at a high height within your home, you should never use a ladder made of aluminum or steel. You will become grounded if there is an electrical surge, and the entire electric current will travel through your body. Instead, you should use a ladder made of bamboo, wood, or fiberglass. Be familiar with the wire code used in your country. Always perform a monthly test on all of the GFCIs in your home. An RCD is also known as a GFCI, which stands for a ground fault circuit interrupter (Residual Current Device). Because they assist in preventing the risk of electrical shock, they have become increasingly widespread in contemporary homes, particularly in moist areas such as the bathroom and the kitchen. It is designed to disconnect on time, which is sufficient to prevent any injury from occurring as a result of an over-current or short-circuit failure.

Hazards
The most common risks associated with electrical work include:

- Harm from contacting live components, electrical hazards, blisters, arcing, fire from electric equipment or connections, detonation due to inappropriate electrical devices, or electrostatic discharge igniting combustible vapors or dust (as in a spray paint booth), and so on.

Electric shocks can also contribute to other harm, for example, by inducing a fall from ladders or scaffolding.

What do you need to do?
Make sure that all potential electrical dangers have been evaluated, including:

- Who they might cause harm to in the future
- How the degree of danger was calculated

- The measures put in place to prevent harm

The risk assessment must consider the surroundings, the method of electrical equipment use, and the type of equipment.

It would be best if you made sure that the wiring and appliances are:

- Adequate for its planned function and the operating environment
- Nothing other than its intended usage

Unfit machinery can become alive in damp environments, contaminating them as well as the equipment itself. All circuit protection devices, including fuses and breakers, must have the appropriate ratings for the circuits they serve. Keep the covers on the isolators and the fuse boxes closed and locked.

All cables, plugs, sockets, and fittings must be sufficiently protected and sturdy for the workplace. Make sure there is a switch or isolator nearby that can be used to quickly turn off the power to the equipment if there is an emergency.

Maintenance

To the extent that it is in your power to do so, you should keep all electrical systems in working order.

Visual inspections of electrical devices, particularly mobile ones, are recommended. Stop using the device immediately and have it checked, fixed, or replaced if:

- There is an issue with the plug or connector.

- The tape has been used to fix the cable, which means it is not secure, that internal wires are exposed, or some other problem.

- There are signs of overheating, such as burns or stains.

A trained professional should do repairs.

Items with a higher potential for damage should undergo more frequent inspections. Equipment with a lower risk of malfunction requires fewer maintenance checkups.

Small, battery-operated gadgets and equipment that run off of a mains-powered adapter rarely require any visual inspection. However, the adaptor that plugs into the wall to power this kind of device needs to be verified visually to ensure it is functioning properly.

Think about whether there should be a more formal inspection or testing of electrical equipment, especially portable appliances. Consider how often this should indeed be done as well.

A pamphlet from the HSE Preventative Maintenance of Electrical Equipment in low-threat settings might inform your decision to conduct routine tests of such devices.

You should schedule regular inspections and tests of your home's fixed wiring installations, including the meter and consumer unit circuits that power your lights, outlets, and any hardwired appliances (like stoves and hair dryers), to ensure that they are in good working order and pose no safety risks. A professional, like an electrician, should usually handle this kind of job.

When is a person qualified to work on electrical systems?
The term "competent person" refers to an individual who has received adequate education, experience, and training for the task at hand.

One technique to demonstrate professional capabilities for common electrical work is to have completed an electrical apprenticeship and gain experience in the field afterward.

Servicing high-voltage switchgear and alterations to control systems are examples of specialized labor that nearly always necessitate prior education and training.

5.2 Protection of Electric Cables

These cables can be found in a wide range of shapes and diameters, allowing them to be used for everything from powering a lamp to generating electricity in a wind turbine. Complex manufacturing methods and highly experienced personnel are required to ensure an electrical cable continues to function reliably over a long period. Coatings for electrical cables serve to insulate them from the elements and prevent mechanical damage. These include insulation, semiconductors, metal screens, filling, sealing, armor, and an outer sheath.

Types of Electric Cable Coating
Insulation

Depending on the desired qualities of the cable, various insulation materials may be employed. The insulating capacity and heat resistance of a material are two of the most important factors in determining its quality. The maximum service voltage of a cable is based on the insulating capacity of the material and the thickness of the cable.

With high-heat-resistance insulation, a given cross-section of a conductor may carry more power than it could with lower-heat-resistance insulation.

An insulating coating is applied to the conductor in the insulation process. The main causes that shorten a cable's lifespan are improper installation, handling, and maintenance, exposure to harsh environmental or climatic conditions, or contact with aggressive substances.

Thermoplastics and thermosetting are the two main categories of insulation. Polyvinyl chloride (PVC), Z1 polyolefins, PE linear polyethylene, PU polyurethane, Teflon (fluorinated), Teftel, and Teflon (non-fluorinated) are some of the most often used thermoplastic insulations in the production of electrical cables.

Semiconductor

Extruded layers of low-resistance insulating materials are used to create semiconductors. After these have been verified as having high thermal stability, the semiconductor elements will do the same.

They have two distinct applications, the first being directly on the conductor of medium and high-voltage cables, and the second between the insulation and the screen.

Top Cable uses the state-of-the-art "triple extrusion" process for manufacturing its semiconductor layers and insulation. A three-layer extrusion head does the actual work. This ensures that no impurities are present because of the direct contact between the semiconductors and the insulator. This prevents partial discharges and cable degradation caused by semiconductor flaws on the inside.

Screen Sade of Metal

The electrical shielding, also known as a screen, prevents any outside interference from corrupting the signals traveling through the connection. Additionally, they protect the power wires to stop any interference with nearby signal circuits.

Filler

The filler is the volume of material used to pack the voids left by the insulated conductors in their wiring. Fabric or synthetic plastic may be used as fillers.

Seat

To prevent the insulation from coming into contact with the metal masses, the SEAT Cables with metal reinforcements are constructed with an extruded layer that goes over the filler.

Armour

Armour, which is a mechanical shield, prevents the cable from being damaged by a variety of potential threats (such as a blow, rodent attack, or traction). Metal strips, wires, or braids make up the armor, which is typically constructed of steel or aluminum.

5.3 The Overload Protection

An overload occurs when the motor consumes more power than it can safely handle. If the motor becomes too hot, the windings could be ruined. To prevent damage to the motor, the branch circuit, and the components in the motor branch, overload protection must be enabled. Overload relays prevent damage caused by overheating by protecting the motor and the motor branch circuit and its components.

The motor starter assembly, which consists of a contactor and an overload relay, also includes an overload relay. Their job is to safeguard the motor by constantly checking the circuit's current. Overload relays have auxiliary contacts that when activated, interrupt the motor coordination circuit and de-energize the contactor if the current exceeds a certain threshold for an extended length of time. The result of this is that the motor loses its supply of energy. Without electricity, the motor and its associated circuitry are protected against overheating and subsequent failure.

How Overload Relay Operate

Because the overload relay is linked in series with the motor, any energy delivered to the motor will also be provided to the relay. This ensures that the relay will always function properly. It will turn off when there is an excessive amount of current flowing through it. When anything like this occurs, the link between the motor and the source of its power is severed. Relays that trip when there is an overload can be reset, whether by hand or automatically, after an amount of time has elapsed. It will be possible to restart the motor once the problem that caused the overloading has been fixed.

Different Varieties of Overload Relays

- **Relay for Bimetallic Overload Protection**

A great number of overload relays are constructed with heater elements, which are also known as bimetallic strips. The bi-metallic strips are constructed using two distinct types of metals, one of which has a coefficient of expansion that is relatively low, and the other of which has a coefficient of expansion that is very high. These bimetallic strips receive their heat from a winding wrapped around the strip itself and act as a conductor of current. Because of the heat, both of the strips of metal will become longer.

The bimetal will bend toward the metal that has a low coefficient of expansion as a result of the dissimilar expansion of the bimetallic strips. Because of the strip's bending, an auxiliary contact mechanism is activated, which results in the normally closed contact on the overload relay being opened. As a direct consequence of this, the circuit containing the contactor coil is broken. Joule's Law of Heating allows us to determine how much heat was produced from a given quantity of energy input. It can be represented by the equation H = I2Rt.

"I" is the overcurrent flowing through the winding of the overload relay, which is wrapped around the bimetal strip.

"R" is the ionic conductivity of the winding all around a thin metal strip.

"t" refers to the amount of time that the current I is allowed to circulate via the winding around the conductive metal strip.

According to the previous equation, the amount of heat generated by the winding will be exactly proportional to the length of time that an overcurrent is allowed to run through the winding. In other words, the operation of the relay is a function of the current squared, so the lower the current, the longer it will take for the overload relay to trip, and the relatively high the current, the faster the overload relay will trip. It will trip much quicker because the relay's procedure is dependent on the square of the current.

When an automatic reset of the circuit is necessary, bimetallic overload relays are frequently specified as the appropriate solution. This occurs as a result of the bimetal having cooled and restored to its initial state (form). When this occurs, the motor will be able to be restarted. If the problem that caused the overload is not resolved, the relay will break again and then reset at regular intervals. When choosing an overload relay, it is essential to exercise caution because frequent tripping and resetting might shorten the mechanical lifespan of the relay and put the motor at risk of being damaged.

In many different applications, the motor is put in a location that maintains a constant ambient temperature, whereas the overload relay and the motor starter may be installed in an area subject to varying ambient temperatures. In these kinds of applications, the point at which the overload relay will trip can change depending on several different circumstances. The temperature of the atmosphere around the motor and the amount of current flowing through the motor are two elements that can contribute to an early motor trip.

These kinds of relays have two different kinds of bi-metal strips: a compensating bi-metal band and a basic non-compensated bi-metal band. The compensated bi-metal strip is the more common form. Both of these strips will bend to the same degree when exposed to ambient temperatures, which will prevent the overload relay from tripping for no reason.

On the other hand, the current flow via the heating element and the motor only affects the principal bi-metal strip. This strip is the only strip that gets affected. When there is an overload, the main bi-metal stripping will activate the trip unit so that the overload can be cleared.

- **Eutectic Overload Relay**

A heater coil, a mechanical mechanism for activating a tripped mechanism, and a eutectic alloy make this sort of overload relay. Eutectic alloys are mixtures of two or more elements that harden or melt at a different temperature.

Overload relays use a tube containing a eutectic alloy, in conjunction with a spring-loaded ratchet wheel, to trigger the tripping mechanism in the event of an overload. The motor's current flows through the heater's little winding. The eutectic alloy tube is warmed by the heater winding throughout the overload. The heat causes the alloy to soften, freeing the ratchet wheel so that it may rotate. After doing so, the overload relay's auxiliary connections will begin to open.

Overload relays that use eutectic technology must be reset manually once they have tripped. This is often done by pressing a reset button located on the relay's housing. The relay's heater unit is selected according to the motor's full-load current.

During overload operations, it is usual practice to use a tube in an overload relay containing the eutectic alloy in conjunction with a spring-loaded ratchet wheel to activate the mechanism that trips the overload. The electricity travels through that one little winding in the motor that controls the heater. It is the job of the heater winding to warm up the eutectic alloy tube when it is overloaded. The molten alloy dislodges the ratchet wheel and makes it possible for it to rotate.

As a direct consequence of the actions taken, the overload relay's closed auxiliary connections will start to open up.

It is necessary to manually reset a eutectic overload relay once it has tripped. For this reason, the lid of the relay will typically feature a reset button that can be used in this situation. A particular heating unit is chosen for the relay to control based on the current drawn by the motor while operating at full load.

- **Overload Relay With Solid-State Protection**

The term "electronic overload relay" is widely used to describe these devices. These electronic overload relays monitor current electronically, as opposed to mechanically, like bimetallic and eutectic overload relays. Even though they come in a wide variety of shapes and sizes, they all offer the same advantages and benefits. One of the best features of these relays is the fact that they don't need a heater. The time and money required for installation are decreased by this design.

Furthermore, the heater-free design is unaffected by the fluctuation in environmental temperatures, which aids in reducing nuisance tripping. Unlike bimetallic or eutectic alloy overload relays, these relays can effectively guard against phase loss. These relays are equipped with a phase-loss detector that activates an auxiliary contact, disabling the motor control circuit. A solid-state relay's overload time and set point can be easily modified.

5.4 How to Protect From Short Circuits?

A short-circuit condition occurs when a circuit enables current to flow through an unintentional path with extremely low electrical resistance. It is the direct contact of two sites with distinct electric potentials.

The short circuit protection system is divided into the following components:

AC Current

- Contact between the phase and the ground
- Contact from phase to neutral
- Contact from one phase to the next
- Contact between electrical machine windings in a phase

Direct Current

- Contact between the pole and the ground
- Interaction of two poles

Damage to conductor insulation, loose, damaged, or stripped wires and cables, and deposition of conducting elements like dust, moisture, and so on can all result in the type of issues mentioned above of connections.

If a current spike equal to a hundred times the working current surges across the circuit, the electrical equipment might be damaged. The destructive effects of short circuits are caused by the two phenomena listed below.

Thermodynamic Phenomenon

When a short-circuiting current runs across an electrical circuit, energy is discharged into the circuit. This thermal action is responsible for the causes of a short circuit:

- Contact melting of conductors
- Insulation damage
- Electrical arc generation
- Thermal components in the bimetallic relay are destroyed.

The Phenomenon of Electrodynamics

When the current crosses, this phenomenon causes intense mechanical stress, resulting in the following conditions:

- Breakage of conductors
- Contact repulsion within contactors
- Conductor distortion in windings

Chapter 6:
The Branch Circuits

A branch circuit is an electrical system component that runs from the main service panel and supplies power to the whole building. Regular outlets and fixtures are powered by 120-volt branch circuits, whereas 240-volt circuits power large appliances.

A branch circuit is a section of an electric circuit that continues beyond the final fuse or circuit breaker. The branch circuits extend from the breaker box to the associated electrical equipment. The last component of the circuit that supplies electricity to gadgets is a branch circuit, which may be categorized in one of two ways: by the kind of loads they support or by the amount of current they can transport.

Branch Circuit and Circuit Breakers

The main circuit breaker controls the main service panel. This acts as the primary disconnect for the power supply to the primary circuit, which acts as the primary disconnect for the power supply to the primary service panel and controls the main service panel. This is typically a two-pole circuit breaker with a rating of 100 to 200 amps that delivers 240 volts of electricity to two hot bus bars that run horizontally across the panel.

Two rows of smaller breakers are located underneath the main one; they are the start of the separate branch circuits that provide electricity to each room in your house. These particular breakers will either be 240-volt breakers that link to two of the 120v bus bars or 120-volt breakers that attach to only one hot dc bus in the panel.

Therefore, your branch circuits will either be 240-volt, which feed big appliances like an electric clothing dryer, an electric stove, and central air conditioning units, or 120-volt circuits, which feed circuits that supply all the regular outlets and lighting circuits.

Circuit Branch Amperage

Branch circuits with 120 or 240 volts may produce varying amounts of power, quantified by amperage. Typically, branch circuits for 120-volt circuits are 15-amp or 20-amp circuits. However, they may sometimes be greater. The amperage is typically 30-, 40-, 50-, or 60-amp for 240-volt circuits.

The lettering on each circuit breaker lever allows you to read the amperage of each branch circuit. Attaching too tiny wires for the circuit amperage provides a clear fire hazard. The wires connected to that circuit must also be able to withstand the load of the branch circuit. Individual wire gauge rates are as follows:

- 14-gauge copper wire at 15 amps

- Copper wire of 12-gauge is used at 20 amps, 10-gauge at 30 amps, 8-gauge at 45 amps, and 6-gauge at 60 amps.

- 2-gauge copper wire at 100 amps

It is not a problem since your home's original circuits were connected properly. However, the new wire must be the correct gauge for the circuit amperage whenever a circuit is expanded. Making the wrong gauge size of wiring is a typical DIY error.

Types

Your house likely has a variety of branch circuit types.

Circuits designated for appliances. These are often necessary by code and only service one appliance. Appliances like electric ovens, dishwashers, freezers, trash disposals, AC, and dryers are served by these 120- or 240-volt circuits. Any device with a motor often needs a separate circuit.

Lighting circuits. These circuits, which provide general illumination for rooms, are exactly what they sound like. A lighting circuit often serves many rooms, and most houses have several. One benefit of isolating the outlet circuits from the lighting circuits is that each room will still have a source of illumination if one circuit is switched off. An electrical bulb that plugs in may be used to light the area while working on the lighting circuit, for instance.

Circuits for outlets. These are merely general-purpose plug-in outlet circuits. They may be unique to a single room or a collection of rooms. For instance, a modest house's second level can contain one or two outlet circuits that service several rooms.

Room circuits. Depending on how the house was wired, a room's lights and outlets could sometimes be serviced by separate circuits depending on the circuit arrangement.

6.1 Luminous Loads

Lighting loads, or the energy required to power electric lights, account for about a third of the energy consumed in US commercial buildings but just 10 to 15% in residential ones. The "Lighting Power Density" of a building, expressed in watts per square foot or square meter, is often used to describe the lighting loads in a structure.

Look at the items' efficiency (or luminous effectiveness) while choosing which lighting products to employ. More energy-efficient light sources and fixtures minimize cooling and lighting loads for the same visual brightness.

Plug Ports

The power required by other equipment, such as computers and appliances, is referred to as a "plug load." In US business buildings, plug loads account for 20–30% of energy loads, whereas in homes, they account for 15%–20% of energy.

Plug loads are sometimes combined with an equipment power density (EPD) and occasionally excluded. Understanding which value you're entering when doing a building analysis is crucial.

6.2 Continuous Load

A continuous load is one whose maximum current is anticipated to last at least three hours. To put it another way, the breaker requires a 25% increase in the continuous load capacity for headroom. Naturally, this calls for a bigger, more costly breaker.

But there is one exception. The extra 25% criterion is waived if the circuit breaker is specified for operation at 100% of its rating. Instead, the requirement is that the device can bear the total of both continuous and noncontinuous loads.

In reality, buying 100%-rated breakers and calling it a day is almost always the right decision. However, as the episode makes clear, it takes a lot of work.

You must perform some load calculations to ascertain whether your loads are largely continuous or noncontinuous. You may size your breakers for 100% of your load if your loads are non-continuous, and you don't need to worry about meeting the 125% criteria. Standard breakers with an 80% rating will be more cost-effective in such a scenario.

Chapter 7:
Grounding of Wiring Devices

The phrase "equipment grounding" and "system grounding" are both often used variations of the term "grounding" in the field of electrical engineering. The interconnection of the earthing system to non-current-carrying electrical components, such as conduit, electrical wires, electrical connections, cabinets, and motor frames, is what is meant by "equipment grounding."

7.1 What is Electrical Grounding

If there is a problem with the wiring system, the current can return to the ground via electrical grounding, which is a backup pathway that gives an alternate route for the current to go. It makes it easier to establish physical contact between the grounding and the electrical devices and appliances in your household.

Electrons moving through metallic circuit wires make up the electricity that flows through a residential wiring system, and this electricity is constantly searching for the most direct path back to the earth to get there as quickly as possible. Therefore, if there is a difficulty with the neutral conductor, anchoring your power wiring will link to the earth and prevent voltage fluctuations that can invite electrical risks. It happens because grounding creates a direct connection to the ground.

How Does It Work?

In an electrical conductor, there is an energized wire that delivers the electricity, a neutral cable that transports that current back, and an "earthing wire" that provides another path for the electrical charge to revert to the ground safely without having caused any hazard to anyone in the occasion of a short circuit. All of these wires are referred as the "live" wires. A ground connection is made by connecting a copper conductor that runs from the metal rod part of the electrical wiring to a set of connections located in the main switchboard.

If the wiring systems employ electrical cables coated in metal, then the metal will typically serve as the ground conductor between both the wall outlets and the control panel. If the electrical cables are not covered in metal, the wiring systems will not need ground conductors. If, on the other hand, the wiring systems employ a cable encased in plastic, then the grounding requires an additional wire. Because electricity is always looking for the quickest path to the earth, the grounding wire is the one that gives a direct channel to the ground if there is any problem when the neutrality wire is damaged or interrupted. Because of this direct physical link, the earth can serve as a path of least resistance, which prevents a device or a person from serving in the shortest way.

Importance of Electrical Grounding

- **It safeguards against potential power surges**

During severe weather, you could be struck by lightning or encounter power outages. A surge of electricity from one of these causes could kill you or fry all of your electronics. By earthing the electrical system, the ground will absorb any stray currents rather than frying any connected devices. In the event of a major power surge, the appliances won't be damaged.

- **It controls voltage fluctuations**

When the electrical system is grounded, it's much simpler to send power where it needs to go. This prevents circuits from becoming overloaded and thus blowing. Any electrical system's voltage sources can be thought of as being anchored to the earth. This contributes to reliable and consistent voltage throughout the grid.

- **Most electrical currents flow easily through the earth**

The earth is a powerful conductor, and it can carry all the surplus electricity with the least resistance, so it is a good idea to ground your electrical gadgets. If you link your electrical system to the earth, any surplus current will flow to the earth rather than through you or your equipment.

- **It saves people's lives and prevents major injury**

If the electrical system is not grounded, the safety of your appliances and maybe your life could be compromised. High voltage can fry and destroy electronic equipment. There is a risk that using too much power could cause a fire, endangering your home and loved ones.

Examining a Current's Grounding Status

You can see if a device is grounded-ready by looking at the manufacturer's instructions. If the appliance has a three-pronged socket and cord, the third wire and receptacle serve as a ground connection between the metal housing and the building's electrical infrastructure.

It is possible to determine whether or not your power circuit is grounded by simply testing the outlets. If the outlet has three prongs, your setup should also have three wires, among which should be used as a ground. Follow the steps below to conduct an electrical grounding test and find out for sure if the current is being grounded or not.

Examining the Grounding of an Electrical System

For an effective electrical grounding test, you can use a receptacle testing device and this 5-point checklist, but only if you proceed with extreme caution.

- Your electrical outlet is the first indication that your home is properly grounded. If the plug has three prongs and a U-shaped slot, it serves as a grounding component.

- Second, plug your circuit tester's red probe into the outlet's smaller hole. Electrical current flows through the "hot" wire located in this plug.

- Third, connect the black probe to the outlet's neutral slot (the larger of the two slots). Doing so will finish the circuit for you.

- Fourth, see if there is a glowing status indicator. If your outlet is grounded, it will turn on, and if not, you need to switch the black and red probes. The outlet is not grounded and should not be used if the indicator does not appear during either grounding test.

- Fifth, to ensure every outlet in your house is properly grounded, repeat procedures 1 through 4 throughout your home. Not all the outlets in older buildings may have been updated during the many renovations that have been done to them.

- By conducting an electric grounding test, you may increase the electrical safety of your current living quarters and rest assured that your electrical installations will remain secure for their useful life.

7.2 Various Household Electrical Wiring Systems

Clamping Wiring

This wiring consists of braided and compounded PVC-insulated or standard VIR wires. Using porcelain cleats with grooves, wood, or plastic, they are attached to walls and ceilings. It is a temporary wiring system, making it inappropriate for residential properties. Additionally, cleat wiring systems are rarely used today.

Advantages

- This fitting can be completed extremely quickly and for a low cost.

- This approach worked quite well for making interim adjustments whenever there was some form of construction going on.

Disadvantages

- This sizing does not appear to be correct.

- A low-quality wire deteriorates extremely fast because it is exposed to the elements and because it is maintained outside, where it is affected by every kind of environment.

Casing and Capping Wiring

Due to the rise in popularity of conduit and encased wiring systems, this method is now considered obsolete. This electrical wiring utilized PVC, VIR, or any other permitted insulated cables. The wires were routed through the wooden casing enclosures, which consisted of a wooden strip with parallel grooves cut along its length to accommodate the cables.

Advantages

- This wiring technique is significantly simpler and more cost-effective than alternative wiring systems.
- This wiring system is highly sturdy and will survive for a very long time.
- This wire is particularly amenable to having its configuration altered.

- Because it completely conceals the wires, there is no possibility of receiving an electric shock.
- When phase and neutral travel in opposite directions, repairs are much simpler to perform.

Disadvantages

If there is a fire in the wires contained within it, then this entire fitting has the potential to catch fire.

Batten wiring

This occurs when a single or group of electrical wires is put across a wooden batten. The wires are attached to the batten using a brass clip and are spaced 10 cm apart for horizontal runs and 15 cm apart for vertical runs.

Advantages

- Wiring is a straightforward and uncomplicated process.
- This way of wiring is more cost-effective than another method of wiring.
- This wiring also has a very nice appearance.
- In addition to that, fixing this wiring is simple.

Disadvantages

- This kind of wiring can't be done out in the open, away from the house.
- Because this wire plays such a significant role in determining the weather, it is not protected from the elements of the outside world.
- It is not possible to utilize heavy wire in this wiring.
- This wiring can only handle up to 220 volts.
- You will need more cords and wires.
- After a long period, cable sag could occur.

Insulated lead wiring

Lead-sheathed wiring employs conductors insulated with VIR and wrapped with an outer sheath of the lead-aluminum alloy containing approximately 95% lead. The metal sheath protects against mechanical damage, moisture, and air corrosion for cables.

Conduit Wiring

There are two conduit wiring styles based on pipe installation:

a) **Surface terminal wiring**

Surface conduit wiring refers to the installation of GI or PVC conduits on walls or roofs. The conduits are fastened to the walls at regular intervals using a 2-hole strap and base clip. Wires are installed within the conduits.

b) **Hidden conductor wiring**

When conduits are placed behind wall slots or chiseled brick walls, the wiring is referred to as concealed conduit wiring. Wires are installed within the conduits. This is popular because it is more durable and aesthetically pleasing.

Advantages

- The use of a conduit wiring system is recommended for both residential and commercial buildings.
- The installation receives the appropriate protection from shock, fire dangers, and mechanical damage as a result of this.
- Cables are guarded against harm from the outside, such as that caused by rodents or short circuits.
- The conduit is tough and sturdy, and it has the potential to last for a very long period.
- Excellent protection thanks to its increased robustness.

Disadvantages

- It is necessary to have the skill to thread the wires and conduit through it.
- Installation requires a lot of work, which is time-consuming and expensive.
- If the cable gets destroyed, cable replacement is particularly challenging in comparison to other tasks.

7.3 The Wiring of Tubes and Knobs

Check the basement of your home for any k&t wire if it was built in 1950 or before. If your home was built later, you won't find any k&t wiring there. A knob and tube wiring system can be identified by the presence of wires routed through porcelain cylinders or "tubes" that are put into holes in the wooden floor joists. You may also come across porcelain "knobs," which serve the dual purpose of securing the wires and preventing them from coming into contact with the wood on which they are routed. In most cases, the wires are insulated using a fabric made of rubberized material.

The absence of a ground wire is one of the most notable characteristics of modern wiring in comparison to the knob-and-tube wiring of the past. Therefore, this sort of wiring cannot accommodate any electrical devices that have plugs with three prongs, and the possibility of electrical shocks and fires is significantly increased. In addition, the black and white wires are not connected in the same way that they are in more contemporary wiring, in which the ground wire, the black wire, and the white wire are all contained within a single cable.

Insulation is yet another point of differentiation between the two. Insulation for knob-and-tube wiring is traditionally made of rubber, but modern wiring uses plastic instead. Knob-and-tube wire typically needs to be changed because its insulation deteriorates with time and causes electrical shorts. It is essential to keep in mind that this is often the consequence of excessive heating or abusive mechanical treatment.

If your house is wired with k&t, getting homeowner's insurance might have proven to be challenging for you. The nature of this sort of wiring does not make it inherently unsafe. Problems happen when the insulation that surrounds the wires begins to deteriorate with age. Both of these scenarios can be avoided by properly maintaining the wiring. Because of the need for open space for ventilation, knob and tube wire should never be routed through insulation, especially blown-in insulation, because this type of wiring must be kept cold. Insulation of any kind surrounding the wiring can potentially cause significant issues. This type of wiring is frequently referred to as "Open Wiring," which helps to underline how critical it is to always keep open airspace surrounding the wire to prevent any overheating from occurring.

How could something possibly go wrong with the knob and tube wiring?

Any one of the issues described further down this page could result in a short circuit or overheating. You might need to upgrade the wiring in your home to prevent such problems. If you are unsure, it is best to get an electrical examination done.

- **Insulation over the wire**

Installing household insulation over knob and tube wiring is extremely dangerous because it creates the potential to start a fire. Insulation made of rubber and fabric is applied over the wiring. It requires a large amount of space to effectively disperse the heat generated by the passage of an electrical current through it. A very hazardous circumstance is brought about when there is no space available since it has been stuffed with insulation.

Knob and tube wire was put in homes at a time when there were very few electrical items in the typical home. This led to excessive use of the wiring. The system can quickly become overheated in modern times because of the prevalence of televisions, sound systems, laptops, washing machines, and dryers. There is frequently an excessive use of extension cords as well as power bars in these situations. The outdated infrastructure just is not built to deal with the surges in electrical demand that occur in today's hyper-connected and digital society. The ground pin, often known as the third prong, must never be removed on power bars and other electrical equipment to make room for the two-prong outlets required by k&t wiring.

- **Adjustments**

The majority of the problems that occur with K&T wiring are due to incorrect alterations made to the wire that was already in place. The fact that it is such an old system means that correct replacement components are not always available. This could be one of the reasons why so many improvised solutions carried out by handymen are so risky. As a result of the ease with which it can be reached in the basement, knob and tube wiring is sometimes spliced in a hazardous manner with modern wire by home handymen rather than by trained electricians, which maybe the explanation for this practice.

- **Damage**

If this type of wiring is damaged in any way, whether by normal wear and tear, repairs performed by a handyman, or other types of damage, it can cause serious problems. Porcelain knobs and tubes are prone to cracking, and cables tend to droop and tear over time, exposing potentially dangerous live wires.

The rubberized fabric insulation that covers k&t wiring eventually gets brittle and can peel off.

When your home is wired with k&t, the presence of an electric lamp or even a television in your living room or bedroom does not constitute a significant safety issue. On the other hand, this kind of ungrounded system could be exceedingly hazardous in settings where there is a chance of coming into touch with water, such as the bathroom or the kitchen.

Chapter 8:
Sizing of Tubing and Boxes

The market is stocked with a wide selection of cables that may be purchased in many different dimensions. Having said that, you are going to need an Electrical Cable Size Calculator to determine which size is going to be suitable for your application. It assists you in understanding the size that is most appropriate for your needs. To get the KW, a power factor of 0.8 is applied to the calculation.

To determine the appropriate size of the cable, we must first divide the voltage carried by the cable by the desired current. For example, if the voltage current of your line is 150 Volts and the aim is 30, you would divide 150 by 30 to get the answer. Using an Electrical Cable Size Calculator helps simplify the process of calculating it for high numbers.

If you are looking for cables for your house and domestic lighting, the standard diameters are either 1.5 millimeters or 1 millimeter. On the other hand, the majority of the time, an electrical cable size of 1 mm is more than sufficient. Only when the cable run will be over a significant distance, and only then, to compensate for supply demand and voltage loss, may a 1.5 be utilized.

When choosing an appropriate cable, a better and more educated selection can be made with the help of an Electrical Cable Sizing Chart. The appropriate size of cable for your application can be determined with the help of these charts. For instance, if a cable with a small diameter is utilized, there is a risk that it will melt because of the significant amount of current flowing through it. Therefore, the Cable Sizing Chart is useful for figuring out both the size and the diameter.

The voltage rating of the Medium Voltage Cable Sizing ranges from 1 kilovolt to 100-volt kilovolts. They feature intricately constructed connections that must be cut correctly. If they are not cut correctly, they have the potential to explode, causing injury to either personnel or equipment. Because of the growing need for different voltage levels, a new concept called Mv Cable Sizing was developed to meet the requirement. The classification evolved in tandem with the rising level of demand. These days, in addition to extra low and extra high ratings, there are now extra extreme classifications.

The Cable Size Calculation Formula is time-consuming and difficult to understand, so we have provided you with the easiest approach to calculate the size suitable for your project. To calculate the size, we use the BS 7671 Cable Sizing technique.

The flow of electricity is met with less resistance when thicker wires are used. They provide a bigger number of charge-carrying electrons as well as a greater number of possible pathways for the electrons to pass through. Therefore, if the voltage remains the same, a thicker cable will carry a greater current. To achieve the desired amount of resistance, it is necessary to select the precise thickness of a cable. The other crucial factors are the size of the wire, which is typically dictated by external needs, and the resistance of the material that the cable is made of.

Divide the voltage passing through the wire by the desired current. If, for example, there will be 120 volts acting on the wire, and you wish to allow 30 amps of current to pass through it: 120 / 30 = 4. In ohms, this is the resistance that you want to achieve as your goal.

Take the length of the cable and multiply the material resistivity it is made of. Copper, for example, has a resistivity of 1.7×10^8-ohm meters at room temperature when measured against an ohm meter. If the required length of your cable is 30,00 meters:

30,000*1.724*10^-8= 0.0005172 ohm sq. m.

Divide the result by the target resistance:

0.0005172/4=0.0001293

This is the minimum required cross-sectional area for the cable.

Simply dividing the area of the wire by pi yields the following value: (0.0001293) / 3.142 = 4.1152 x 10-5.

Determine the answer's square root as follows: (4.1152 x 10^-5) ^ 0.5 = 0.006415. The distance between two points along the wire, in meters.

To convert your response into inches, multiply it by 39.37. 0.006415 x 39.37 = 0.2526.

Divide the result by 2, as follows: 0.2526 multiplied by two equals 0.5052 inches. This is the minimum thickness that the cable can have. It is very similar to the typical 16-gauge wire in appearance.

8.1 How to Check the Conductivity of Cable

The conductivity of electricity is a physical attribute that describes how well one substance conducts electricity in comparison to another. When electrical charges move in response to a difference in potential electrical energy, a current is produced. The term "conductivity" refers to the relationship between the density of this current and the intensity of the electric field. It is possible to calculate the electrical conductivity of a test material by measuring the resistance, area, and length of the material. To facilitate the measuring process, the test material is often shaped like a box.

For higher precision, an ohmmeter with four terminals should be used. Because one pair of these ohmmeter's terminals measures current and the other pair measures voltage, this particular form of ohmmeter has a higher degree of accuracy. Because of this, the ohmmeter can disregard the resistance presented by the initial pair of terminals.

Take down the reading for the test material's resistance. The computation is performed automatically by the ohmmeter. R equals V divided by I, where R represents the resistance in ohms, V is voltage in volts, and I is current in amps.

Take readings in meters for each of the dimensions of the test material. The length is equal to the distance between the two terminals of the ohmmeter. The ohmmeter measures current flow across a specific area of the surface, and this is referred to as the area.

Determine the electrical conductivity of the material by calculating it using the resistance, the length, and the area of the current. The formula for determining the resistivity is as follows: p = RA/l, where p represents the resistivity, R represents the resistance, A represents the area, and l represents the length. The formula for determining conductivity is s = 1/p, with s standing for the conductivity. Because of this, the conductivity can be written as s = l/AR, and it will be measured in ohm meters-1, which is also referred to as Siemens.

8.2 Electrical Cables

Electrical power can be transmitted through an electrical cable, often known as a power cable. Electrical cables are used to establish connections, which are required for the proper operation of electrical infrastructure and electrically powered equipment, including power plants, wired computer connections, tv, and cellphones. There is a wide variety of electrical cables, each with its unique configuration, size, and performance characteristics.

Components that make up electrical wires and cables

A minimum of two conduction paths and a protective outer jacket are components shared by all types of electrical cables. Each one of the conducting wires contained within the outer protective jacket of medium to high power cables that transport high voltages can be individually sheathed in insulation. Copper is the most frequent metal used in the construction of electrical conductors. The outer jacket and the protective and insulation layer are both made of synthetic polymers.

- **Coaxial cable**

A dielectric insulator is present in a coaxial electrical cable. The insulating layer is encircled by a copper-woven barrier, and then an outer plastic sheath is twisted around the whole thing to finish it off. The size, performance, flexibility, capacity to handle power and the cost of coaxial cables can vary greatly. Home audio and video equipment, television networks, and the various nodes that make up a local area network can all be connected with their help. There are several varieties of coaxial cables, including hard-line, RG/6, and semi-rigid cables.

- **Ribbon cable**

Multiple insulated wires are laid out in parallel to one another to create electrical cables. This type of cable is also referred to as flat twin cables. The transmission of numerous signals and data can take place simultaneously thanks to these parallel lines. According to the book "Optical Communications Essentials," the average ribbon cable has anywhere from four to twelve wires inside of it. It is frequently employed in the process of connecting various network devices. In computers, the motherboard is also connected to the various core components of the central processor unit (CPU) using ribbon cables.

- **Twisted pair cable**

Electrical cables known as "twisted pairs" consist of pairs of insulated, color-coded copper wires twisted to each other. In various types of twisted pair cables, the diameter of each wire can range anywhere from 0.4 to 0.8 millimeters, and the number of pairs can change as well. The cable's resistance to cross-talk and external noise will be proportional to the number of pairs it contains; the bigger the number of pairs, the better. Twisted pair wires are not only flexible but also easy to install and not very expensive. They are utilized in the process of telephone cabling as well as the wiring of area networks.

- **Shielded cable**

One or more insulated wires form the core of a shielded electrical cable, which is then sheathed in either woven braid shielding. Because the cable is shielded, it is protected from radio and power enhancement from the outside, which enables the signal transmission to proceed without any problems. Shielding is typically utilized for high-voltage power connections.

8.3 Device Box Sizing

To avoid damage to the conducting insulation, pull panels, junction boxes, handhole casings, and tube bodies have specific dimensions that must be adhered to. These standards apply to all wires with a gauge size of four American wire gauges or more. To demonstrate how these requirements avoid damage to the conductor insulation, let's take two extreme cases of a scenario known as a straight pull.

Let's say you have a raceway that measures 2 inches that goes inside a box that measures 10 inches square and another raceway that measures 2 inches that goes out of it. How are you going to drag many 1/0 AWG wires through the box without bending them so much that you break the insulation? If you're pulling them through one side of the box, how are you going to draw them into the other side? Answer: You're not.

Imagine now that you have those two raceways placed on top of a box four feet tall. Since it is obvious that you have a sufficient amount of room, you won't need to make any significant turns in the wires. Calculating the size of the required box is necessary since you do not have the luxury of erecting 4-foot boxes for every wire pull that you perform.

The situation we have just gone over is known as a straight pull. In this scenario, the conductor enters the box on one side and departs on the other; however, there is neither an angle nor any splices involved in the process. This installation is what's known as a "straight-through."

Straight pulls

Why would you put in a pull box if you were just going to use a straight pull? Why even bother to divide up the raceway with a box? A very lengthy conductor line is one reason; another one is to restart the 360-degree bend constraint of the corresponding raceway article. The pull box not only offers an additional lubrication point but also lessens the force required to pull that particular line in the first place.

The length of the box used for straight pulls must be at least eight times greater than the trade size of the raceway used.

U-pulls, splices and angle pulls

There are additional uses for junction boxes outside, simply simplifying the process of doing a straight pull. You can use a junction box to change the direction of the raceways, splice conductors, combine runs from multiple raceways into one raceway, and split runs from one racetrack into multiple raceways. In circumstances like these, the sizing requirements for junction boxes become somewhat more problematic. As in the straight pull, begin with the largest raceway size possible, but in addition, complete the following steps:

1. It is more accurate to divide the raceway size by six rather than multiplying it by eight.

2. The sizes of any other raceways located on the same wall and row should be added together.

3. Combine the findings from the first and second steps.

4. You must ensure the length of the box is at least equal to the number you determined in step 3.

Calculate each row of raceways individually in areas where there are many rows of raceways. Then, to determine the size of the smallest possible box, use the row containing the most distance.

How much space is required between the two raceways when conductors enter a junction box through one raceway and depart the box through another raceway (that is not a straight pull) of a different size? You will require a trade area six times larger than the larger raceway. Calculate the distance not from one object's center to another's center but from the nearest edge of one to the nearest edge of the other.

8.4 Tubing Sizing

A conduit pipe, commonly known as an electrical conduit, is a strong tube for electrical wiring. The primary goal is to shield the wire from damage in areas where it is most at risk. This is why it is commonly used in less obvious places like cellars, crawlspaces, and lofts.

A conduit pipe may remind you of an armored cable. The individual wires in an HVAC system are protected like an armored cable.

Cable management systems employ a wide variety of hardware in addition to the electrical conduit. A few examples of these parts are cable carriers, wire channels, and ducts for securing wiring. Typically, conduits are made from metal, fiber, plastic, or baked clay. These substances may be rigid or malleable.

Imagine you're working on a project requiring you to lay ductwork or electrical conduits. The United States National Electrical Code and/or your local building regulations are excellent resources to look to in this situation, as they detail proper methods for installing, constructing, and inspecting electrical systems.

Various forms of wire and cable for electricity

The conduits can be distinguished or categorized according to the tubing material and wall thickness. Conduits are made from various materials, and the chosen type depends on their intended use.

- **Electrical metallic tubing**

Electrical metallic tubing (EMT) is non-flexible tubing with threaded fittings and couplings composed of galvanized steel or aluminum. These are secured with a set screw or other compression-type fastener.

You needn't worry about cost or accessibility, as treated fixtures are readily available and relatively inexpensive. Although they appear on the list, tubes are not conduits but rather a different category entirely.

Thin-walled Electrical Metallic Tubing (EMT), is utilized in places where there will be no stress on the conduit.

They are rather rigid but may be bent with a tube bender. Wiring in residential or light commercial buildings commonly uses EMT because it is safe for use in enclosed spaces. Therefore, emergency medical teams must build their equipment with special seals to prevent water damage.

There are EMTs available with a protective coating that prevents them from rusting, allowing them to be used indoors and out.

- **Flexible metallic conduit**

A flexible metal conduit is termed 'Greenfield' after the scientist Edward Greenfield, who devised the metallic conduit invention. Steel straps are coiled with self-locking mechanisms to create a tube thru which wires can be fed.

The metal strips and interlocking construction give both rigidity and adaptability, while the high-quality aluminum alloy ensures strength and longevity.

FMC is commonly utilized for the final few feet of wiring because, in contrast to EMTs, it can be bent in a situation where normal conduit would be difficult to work with.

Walls and other structures benefit from their spiral design and widespread adoption. Besides the flexibility, FMC has other advantages, such as dampening vibrations and permitting mobility.

This is why you will notice FMC in-house wiring for compressors, motors, and industrial machinery. Unlike other tubes, FMC components are simple to swap out, replace, and upgrade.

The biggest disadvantage of FMC is that it is not highly corrosive and doesn't shield the wires against impact. Since it is unsuitable for burial or embedding in concrete, it is utilized primarily indoors and only sometimes outside.

- **Liquid-tight flexible metallic conduit**

In the same way that FMCs are manufactured, so too is a liquid-tight flexible metal conduit, sometimes called "seal tight." The main distinction is that its exterior is made of plastic or non-metallic material.

Unlike FMCs, they may be used in wet and dry environments thanks to their waterproofing, crack resistance, and UV resistance. The joint and end connectors must be watertight if they are to be utilized in a damp environment.

Low-frequency motor controllers (LFMCs) are commonly found in outdoor appliances like HVAC systems (AC). However, despite LFMC's durability in wet environments, it can be easily damaged by impact. Therefore, it should not be subjected to forces or enclosed in concrete.

- **Rigid PVC conduit**

Among the many varieties of wire used in commercial and industrial structures, rigid polyvinyl chloride is common. It is employed in regions exposed to fluids or moisture and functions similarly to plastic-vinyl chloride (PVC) plumbing pipes.

Like their PVC counterparts, RPC plumbing pipes are waterproof and can have plastic fittings cemented onto them for installation. The primary distinction is that RPC is resistant to most acids.

Due to its PVC construction, this conduit does not suffer from the effects of rust and corrosion when buried below.

- **Rigid metal conduit**

Rigid metal tubing is frequently used in commercial structures. Strong galvanized steel, stainless steel, or aluminum with a threaded fitting is typical.

A rigid metal conduit, or RMC, is an aluminum conduit with extremely thick and sturdy walls that shield the wires inside from damage in the event of drops, knocks, or other mishaps. However, this indicates that they could be more convenient, inflexible, and costly.

You'll probably need to use pressure fittings to snip them, but we recommend getting expert help if you need more confidence in your cutting abilities. Aluminum RMCs are very resistant to moisture, corrosion, and elements.

But since aluminum reacts badly with concrete, they need to work better with that material. However, nowadays, you can buy RMC threaded and coated with a specific coating so that it may be implanted in concrete.

Galvanized steel (GRC) RMCs are versatile and can be utilized in various settings; for instance, as a grounding conductor for electrical appliances. They also serve as a sturdy base for wiring and control panels.

Intermediate metal conduit (IMC) is a thinner and lighter alternative to RMC. It serves the same function and is used for the same purposes as RMC, but LMC is preferred in modern buildings because of its lower weight.

- **Electric non-metallic tubing**

Non-metallic electrical tubing is often a flexible corrugated plastic tube with bonded plastic fittings for assembly. It is flame-retardant, which may suppress or delay the spread of fire, and resistant to water and moisture. These tubes are installed indoors, typically inside walls, and are protected from the elements. They are highly flexible, allowing for in-block installation. Yet, they must not be buried in a shallow grave.

Chapter 9:
The AC Load Calculation

According to the definition provided by the NEC, a housing unit is a single building that contains all of the facilities necessary for an individual to live independently.

9.1 Single-Family Dwelling

When it comes to loading estimates, dwelling units have pretty specific requirements. Although the majority of the actual load assessment criteria are located in Art. 220, others are dispersed throughout the Code and continue to be relevant when specific calculations are being performed. When performing calculations for housing units, make sure to keep the following points in mind:

- Voltages. Calculate loads for branch circuits, feeder circuits, and service circuits using the average voltage level [220.5(A)] unless other voltages are expressly stated. The nominal voltage for a solitary dwelling unit is commonly 120/240V, but it can sometimes be higher.

- Motor VA. Use the voltage and current numbers from the motor table instead of 120V, 240V, or 480V. Some examples of these values include 115V, 230V, or 460V. By using the motor's rated current and voltage ratings which were utilized in the development of the Code Tables, we can obtain a significantly more accurate rating for the motor's VA.

- Rounding. You can ignore the fraction [220.5(B)] whenever the results of your calculations produce a fraction smaller than 0.50A.

- Receptacles. As far as there is more than one socket on the circuit, you are free to use either 15A or 20A receptacles even though the circuit is rated for 20A. According to

[210.21(B)(3)], a dual receptacle is the same as two regular receptacles when it comes to these considerations.

- Constant amounts of work. According to the definition given in Art. 100, a continuous load is one in which it is anticipated that the maximum current will remain for at least three hours continuously. One illustration of a constant load is a permanently installed electric heating system [424.3(B)]. Increase the capacity by 125% to determine the appropriate size of the switchgear wires and overcurrent protections for a constant load.

- Laundry rooms. A receptacle in the laundry area is necessary [210.52(F)], and 210.50(C) stipulates that at least one of these receptacles must be located within 6 feet of a washing machine. [210.8(A)(7)] requires that any receptacle within 6 feet of the outer parts of a laundry basin must have GFCI protection.

The circuits needed

A residential unit is required to contain the following circuits in addition to the circuits needed to serve specialized appliances, as well as the circuits that are needed to serve the ambient light and receptacle load:

A minimum of two branch circuits for small appliances rated at 20 amps and 120 volts each should be installed for any receptacles located in the kitchenette, dining hall, reception room, pantry, or other places used for eating [220.11(C)(1)]. [210.52(B)(2) Ex] It is strictly forbidden to utilize these circuits to power other receptacles, such as lighting channels or outlets located in other portions of the building. These circuits are factored into the calculation for the feeders and services at a total of 1,500 VA for each route [220.52(A)].

It is one branch circuit of 20A and 120V for the receptacle for the washing room(s). Only the receptacle connections in the laundry area can be served by it [210.52(F) and 210.11(C)(2)]. It cannot service any other outlet, such as illumination.

9.2 Calculations of the Feeder and Service Levels

Under typical living conditions, occupants do not utilize all loads at the same time; hence, "demand factors" can be applied to several of the dwelling unit loads to determine the appropriate amount of service. Some of the demand factors supplied in the Code are only meant to be used in residential settings, while others are only permitted in non-residential settings.

Because of this, you need to be very careful to only apply demand factors in the ways permitted by the NEC.

The National Electrical Code (NEC) specifies two techniques for calculating the residential service load: the standard approach and the optional method.

The calculation for feeder and service loads using the standard technique

Load on the VA general lighting system

Include a minimum of 3 VA per square foot for general illumination and general-use receptacles when computing the loads for branch circuits, feeders, and services in residential properties [220.12]. When calculating the area, you should use the dimensions of the house taken from the exterior. Avoid including open terraces, garages, or other spaces that cannot be converted to accommodate different uses in the future.

Circuits for home electronics and laundromat equipment

The general illumination and all 15A and 20A, 125V broad sense outlets are included in the 3VA per square foot guideline.

However, this rule does not include receptacles for small appliances and laundry circuits. As a result, you need to compute those based on 1,500 VA for each circuit. See 220.14(J) for instructions.

The total amount of branch circuits

The load of the general lighting and the rating of the circuits should be used to figure out the necessary number of branch circuits for light sources and receptacles used for general purposes.

9.3 Electrical Load Capacity and Calculations

It is important not to treat the installation of electric cables in your house lightly. You must understand very well the maximum electrical load of your home if you are considering either moving into a new house or improving your current residence.

If you can plan and control the power distribution throughout your property, it will help if you have a good understanding of the load capacity. If there is a requirement, an experienced electrician can assist with the installation of new or improved electrical systems.

What does the electrical load capacity mean?

The maximum amount of current that may be carried by a wire is referred to as its electrical load capacity. To ensure that all of the electrical appliances and equipment in a home are powered without incident, the residential electrical system must have a specified load capacity. If your home has power outages regularly, you should calculate the total wattage of all of the appliances and electronics in your home to ensure that the total doesn't exceed the maximum capacity of your wiring.

How can you tell if the electrical service you are currently receiving is sufficient? Amperage, also written as amperes, is the unit of measurement for the overall electrical capacity of an electrical service. When building large electrical improvements to your home, it is important to determine the total load that will be placed on the electrical system by factoring in everything that needs the power to function.

Wattage or watt-hours are the units of measurement for the electrical load. Your home's wiring is designed to carry a certain total quantity of electric current at one time before it becomes dangerous. This is referred to as the electrical load capacity.

The national electrical codes serve as the foundation for determining whether or not wiring is secure and satisfies the safety criteria for a specific place. According to the building code, the individual branch circuits in a contemporary home should have a capacity of at least 40 amps, which is equivalent to eight outlets with a standard rating of 15 amps each.

9.4 How is the electrical load capacity found?

The ampere load of your home's appliances and lights determines how much electricity your home uses. Check the labels of all the appliances and gadgets plugged into the breaker box to determine the total amount of watts (W) they consume.

It might be challenging to determine a property's total wattage, but a few things can be checked. Ideally, you would check the wattage of each appliance and gadget, but this is only sometimes practical or practicable. In these cases, you'll have to estimate the typical wattage of items in a particular category.

A special cord termed an electric load computation cord can be purchased at hardware stores and used to determine the maximum electrical load your home can handle without knowing the wattage of your appliances.

Start by multiplying the amperage displayed on the load calculation cord by 120 to get the voltage, and then add the amperage of each device you intend to use on that circuit. When the sum exceeds the circuit's limits, you must unplug some appliances so that the circuit can supply electricity to your home without overheating.

Water flowing from a pipe is an analogy for how the circuit operates. Volts represent water pressure, while amps indicate water flow rate. Power output in watts is symbolized by a water wheel, where the value of this quantity is proportional to the surrounding pressure. The calculations provided here can be used to determine the capacity of individual circuits, the electrical loads on those circuits, and the total electrical service panel.

9.5 How Is It Calculated?

It's possible to calculate the amperage service required by your home with a little bit of math. For instance, a 100-amp service is suitable for a home of less than 3,000 sq ft without air conditioning and electric heat, while a 60-amp service is woefully inadequate for a modern home.

A 200-amp service may be necessary for a home with central air conditioning and electric heating greater than 2,000 square feet. The general branch circuit breakers and one or two household appliances in a modest to medium-sized residence can be safely powered by a 100-amp service.

If your home has less than 2,500 square feet and you heat it using gas appliances, this may be enough service for you. Households that rely solely on electric appliances and HVAC systems may consider upgrading to a 400-amp service.

Chapter 10:
Tips and Tricks to Better Study

10.1 Tips So You Pass the First Time

Practice with previous tests

Exam practice is one of the most efficient strategies for preparing for tests; this aids in becoming used to the structure of the questions.

Describe your responses to others

It will help you get clarity.

Schedule your exam day

Don't wait until the test day to discover you don't know how to get there or what to bring; make sure everything is ready well in advance. Run the journey through a test run if you can.

Graphs and flowcharts are useful

Visual aids can be quite helpful for review. It is helpful to start by listing everything you know about a matter at the outset and then figure out where your knowledge is lacking.

10.2 Tips to Pass Your Journeyman Electrician Exam

Read the text all the way through

Complete a thorough reading of the material while you are starting so you can begin understanding the information. Giving oneself a solid foundation to build on is crucial for success since repetition is crucial for learning.

Take notes by hand

According to studies, taking well-organized handwritten notes facilitates learning. Because you tend to organize knowledge into more practical forms while taking handwritten notes, they have been demonstrated to be more productive. It creates a greater feeling of knowledge and more intense mental engagement.

Make a study guide with the ideas you need to review

You may link ideas by creating a study guide or "cheat sheet," which also helps you see trends and keep track of the material you need to review. A study guide will serve as a direction for your future study and increase the effectiveness and efficiency of the learning process.

Give topics that you need to understand as much priority

Humans may desire to learn things we already understand to feel better about themselves. Spend a lot of time studying new ideas or refining ones you need help comprehending to succeed.

Set a target

Setting daily goals for your study time will hold you responsible, keep you on track, and stop procrastinating. For instance, to ensure that you are actively progressing in your learning, establish a target of studying 25 questions today.

Follow the 50 – 10 Rule

According to the 50 – 10 rule, you should study for 50 minutes before taking a 10-minute break. This action keeps your mind active and avoids mental fatigue. Always stretch and maintain proper posture!

Link notions and theories to actual experiences

Your knowledge and comprehension of the theory or idea are solidified when you can relate it to real-world applications. It enhances memory and may increase productivity in your studies and professional lives! Try to relate what you learn in class to your everyday tasks at work, for instance, if you are an apprentice at a construction site.

Test yourself or allow someone else to do so

By testing your knowledge and increasing repetition, quizzes let you assess what you have learned. You may then decide what needs to be improved and go on from there.

Use practice exams

Practical examinations allow you to better test your knowledge and comprehension, as quizzes provide you with a true reflection of your knowledge.

Create a test environment simulation

Simulating the examination atmosphere is crucial while preparing for or taking a practice test. Make sure to take at least three complete practice examinations to prepare yourself for the test.

Attempt to minimize disruptions

Home study may be challenging. To make the most out of your study time, try to be attentive and involved in what you are doing. Distractions from the visual and aural world might prevent you from learning. Close the door, mute your phone, and think about inserting earplugs.

Keep your study area organized and spotless

It's crucial to be calm and undistracted when studying. Remember, a clean environment promotes a clear mind!

When in doubt, seek assistance

If you don't know the solution, don't give up! Never hesitate to ask for assistance from your lecturers, other students, or even your boss! Try an internet forum or get advice from a professional in your industry if you need help figuring out the solution or understanding why something is the correct response.

Do not study when you are sleep-deprived

Your brain may learn that studying when fatigued is equivalent to taking a sleep. You're more prone to experience test fatigue if you study when exhausted. Make sure you get enough sleep and study while you're awake.

When you're most alert, study

It all depends on the individual. Do you like the mornings? Do you like sleeping in late? When you are alert, you should study. You'll be able to concentrate for longer periods and remember more information. Both day and night study times have advantages. It takes less artificial illumination to study throughout the day. Sunlight improves your eye's health and awakens you. However, studying at night is often more tranquil, calm, and laid back. Many believe studying after work allows them to think "beyond the box."

Explain the exam concepts to a person who is unfamiliar with the subject

Teaching someone else is a tried-and-true method of ensuring that you comprehend a subject. It may be useful to "dumb down" complicated subjects for someone else before getting feedback on any misunderstandings to identify subjects you may not fully comprehend. If you cannot respond to their inquiries, you may need to go over that subject again.

Speak with a test-passing individual

You may get valuable insight into whether you are preparing for the exam properly or need to attempt a different approach by speaking with someone familiar with the test's difficulty level. Hearing about someone else's test and study experiences will help you develop new strategies.

Consume brain food

Your memory and focus are significantly impacted by food. While studying, and on test day, make healthy food choices such as yogurt, fresh fruit, almonds, salmon, and nuts. The energy from these meals is released gradually throughout the day. Conversely, sugar can make it harder to concentrate and cause your energy levels to drop within an hour.

10.3 Causes of Test Anxiety

Although we often think of worry as an uncontrolled mental condition, several straightforward, everyday factors might trigger it. Not feeling sufficiently prepared for a test is one of the most frequent reasons for test anxiety. Time management is the most frequent cause of this sensation, while other factors like bad study habits or a lack of organization may also contribute. You will not be properly prepared for the exam if you start studying too late, plan your study time to cover everything, or become side-tracked while studying. This might result in cramming the night before, leaving you physically and psychologically spent for the exam. Being unprepared and unsure of what to do about it may lead to emotions of worry, dread, and despair. This is also a result of poor time management.

Sometimes, exam anxiety results from unresolved dread rather than preparedness for the test. This might be a former test failure or subpar test performance in general. It could result from comparing oneself to someone who seems to be doing better or from the pressure of meeting standards. Fears about how failing this exam may affect your educational and professional ambitions could be the cause of anxiety. Even though these anxieties are sometimes utterly unfounded, they may hurt your test performance.

10.4 How Can You Reduce Test Anxiety?

Although exam anxiety is a significant issue, the good news is that it is treatable. It doesn't have to be in charge of how you think and retain knowledge. You may start taking action now to overcome anxiety, even if it could take some time.

It's acceptable to feel anxious before an exam; it helps you concentrate and focus. However, when you have test anxiety, your performance on the exam may suffer, and you may feel awful. Anyone who must take exams for job progress or certification, whether in elementary or secondary school, college, or employment, might have test anxiety.

Here are some techniques to assist you in coping with exam anxiety:

Learn effective study techniques. Study skills programs and other tools at your school may help you improve your study habits and test-taking strategies. Exam anxiety can be reduced with careful review and practice of the material.

Create a reliable pre-testing schedule. Find out what works for you and use the same procedure each time you prepare for an exam. Your stress level will decrease, and you'll be more prepared.

Speak to your instructor. Make sure you are aware of the topics that will be covered on each exam and how to prepare. Additionally, let your instructor know that you have exam anxiety. They could have advice that will enable you to succeed.

Study relaxation methods. You can help yourself remain calm and assured before the exam by practicing relaxation techniques such as breathing techniques, releasing each muscle one at a moment, or closing your eyes and envisioning a successful outcome. These techniques can be used at any point during the exam.

Remember to eat and drink. To operate, your brain requires energy. On exam day, eat something and drink plenty of water. Avoid consuming caffeinated liquids such as energy drinks or coffee, which may heighten anxiety, and sugary drinks like soda pop, which can cause your blood sugar to spike and then plummet.

Take a workout. Exercise may help you relax on test days and regularly.

Get plenty of rest. Academic achievement and sleep are intimately correlated. Teenagers and preteens, in particular, need frequent restorative sleep. However, people also need a restful night's sleep to function at their best at work.

If needed, see a licensed professional counselor. Working through emotions, beliefs, and behaviors that contribute to or exacerbate anxiety may be accomplished via talk therapy with a psychologist.

10.5 The Best Tip for Taking the Written Examination

Here's a suggestion for answering multiple-choice questions: attempt to come up with a response to each question before you consider the other options.

On some multiple-choice tests, you cannot change an answer once you have entered it. This is because a scanning device has marked them. As a result, always verify your work twice before writing an answer down.

Move on from a question if you don't know the answer. Later, you may revisit it. Make an informed estimate based on your knowledge of the situation if you need clarification on the answer. Guessing is a better final choice if you don't know than leaving the question unanswered.

Utilize the method of elimination. Once you have a solution, eliminate any possible answers you believe to be erroneous.

Disbelieve test-taking myths such as the idea that the longest response is more likely to be accurate or that the response is always "C."

Keep to your first-choice response if you are hesitating. Usually, the first response that comes to mind is the right one.

Chapter 11:
Journeyman Electrician Exam Simulation Test

Quiz Number 1

1. At any given location, what is the utmost percentage of the cross-section of a metallic wireway that can be covered by splices, taps, and circuits?

2. The full-load operating current specification for a 480-volt, 3-phase coiled-rotor ac motor with 30 horsepower has the same value as _____.

3. Which is the UTMOST number of twists that a single line of subsurface constructed stiff (PVC) that is 50 feet in width across pulling locations is allowed to have?

4. From dwelling units, everything 125-volt, separate-phase, 15- or 20-ampere container channels that are set up inside a radius of six feet of the exterior perimeter of a fall in the laundry area are required to be stipulated with _____.

5. Standard commercial size stiff Category 40 (PVC), which is 2 inches in diameter, should be braced at _____.

6. Although the breakers are specified and labeled to indicate alternatively, conducting materials of dimensions 14 AWG through 1 AWG must be assumed to have a rating for the temperature of ____ while they are being terminated on the breakers. This is the case even if they have a smaller gauge.

7. Within the confines of a Class FS box with a single gang and a depth of one inch, a maximum of _____ twelve-gauge AWG wires are allowed.

8. Where unattached conductor cables of measurements 1/0 AWG using 4/0 AWG have been placed in ladder-type cable compartments in manufacturing facilities, the HIGHEST acceptable tolled distance for the cable containers shall be _____.

9. Pipes or tunnels used to construct grounded electrodes CANNOT be any smaller than _____ electrical commerce standard.

10. What is the smallest size of copper SE wire with grade THHW insulating that could be utilized as unsupported (phase) connection-entrance cables for a 120/240-volt, single-phase home service that is certified for 150 amps? What exactly is the maximum voltage that can be carried by these conductors?

11. When wires from multiple systems are put in an identical raceway, the color white or gray must be used for the grounded circuit of one of the networks. The additional grounding component of the system is required to be _____.

12. The magnitude of section-circuit connections that feed a plurality of motors must be at least _____ of the filled with a bunch-load current capacity of the most highly rated motor in the entire group, and it must be a minimum of 100 percent of the filled with a bunch-load present capacity of the remaining motor(s) in the same group. This is required by the National Electrical Code.

13. In situations in which exclusions are not going to be taken into account, service conductors that will be put in place as open conductors need to have a distance of NOT Short THAN between themselves and any openings, porches, entrances, or terraces that are planned to be accessed.

14. What is the absolute smallest size of the machinery grounded conductor that must be used for a 5 horsepower, three-phase, 208-volt motor that has overload safeguards valued at 20 amps and branched circuit, short-circuit, and surface-fault prevention valued at 30 amps each?

15. Where the full-load current limit of a separate-phase, 240-volt cooling unit appears on the license plate frame as 35 amps, what is the MINIMUM number of 75°C conductors made of copper needed to power the cooling device?

Answer Key of Quiz Number 1

1. Ans: 75 percent

2. Ans: 40 amperes

3. Ans: 90 degrees

4. Ans: GFCI protection and AFCI protection

5. Ans: 5 feet

6. Ans: 60°C

7. Ans: 6

8. Ans: 9 inches

9. Ans: 3/4 in.

10. Ans: 1 AWG

11. Ans: green

12. Ans: 125 percent

13. Ans: 3 feet

14. Ans: 10 AWG

15. Ans: 8 AWG

Quiz Number 2

1. When building wiring for hazardous electronic devices, the Highest voltage that can be applied to the grounding terminal must be at least _____.

2. Branch-circuit wires that serve an electric water heater of a permanent storage kind with an annual consumption of 120 gallons or less must have a magnitude that is NOT below ____ of the water heater's filled with a bunch-load current. This is required by the National Electrical Code.

3. External open wires that are not more than 1000 volts are required to have a distance from the ground of NOT below the following from designations, fireplaces, and television antennae.

4. Domestic appliances with surface heating components that have a MAXIMUM consumption of more than ____ should have their power supply split into a pair or more circuitry. This is required by the National Electrical Code.

5. On the power side of each fuse in an electrical system that has a polarity of not less than _____ to ground, a disconnect method is required to be installed.

6. When wires are put in a pipe or tubing nozzle that is _____ in size or fewer hours, the magnitude modification factors that involve a total of three (3) current-carrying circuits in the track do not need to be performed. It happens because the length of the conduits or tubular nipples is shorter than the size of the track.

7. If a barrier is held up by an overhead ceiling framework, then it must be attached to the framing using one of the following techniques:

a.   The screws

b.   Bolts

c.   Rivets bolts

d.   Each of the preceding options

8. The temporary wire must be disconnected within which of the subsequent period limits after construction is finished or once the purpose for which it was placed has been fulfilled:

a.   31 days following an occurrence

b.   Within sixty days following the end of an event or building project

c.   Approximately ninety days after the finish

d.   Immediately

9. Which of these options will offer extra safety precautions if excessive quantities of material in or on the structure get electrified and might potentially be vulnerable to touch by individuals?

a.   Proper bonding and grounding procedures

b.   Bonding

c.   A connecting jumper

d.   None of the preceding options

10. Conducting materials that feed a fire pump motor need to have a capacity that is not less than ___ % of the full-load current drawn by the fire pump machine and ___ percentage of the electricity used by any connected fire pump auxiliary the machinery:

a.   60, 100

b.   125, 100

c.   105, 115

d.   110, 100

11. Under which of the subsequent situations is it permissible to enable a moving cable to operate without using a raceway?

a.   The distance between the first point of support and the starting point cannot be more than six feet.

b.   The electrical conductors are bundled up with one another and secured with tape.

c.   The conducting instruments are still sheathed in their first covering.

d.   Each of the preceding options.

12. In which of the subsequent systems is the use of rigid metal conduit allowed?

a.   Any forms of tenancy

b.   Corrosive environments

c.   Wet locations

d.   Each of the preceding options

13. Which of the subsequent applications does not allow the use of concealed wiring using knob-and-tube connectors?

a.   Commercial garages

b.   Film production companies and studios

c.   Dangerous or otherwise categorized areas

d.   Each of the preceding options

14. Which of these tasks cannot be accomplished with the help of a cable bus:

a.   Circuits that serve branches

b.   Feeders

c.   Cables for the service

d.   None of the preceding options

15. Which of the aforementioned terms best describes an insulated conductor arrangement that also includes connections and conductor expulsions and that is entirely contained within a vented protecting metal housing?

a.   Busway

b.   Busbar

c.   Cable bus

d.   A raceway

16. It is permissible for branches that originate from busways to utilize which of the subsequent cabling approaches:

a.   The cable of type MC and the cable of type AC

b.   Conduits of types RMC and FMC, respectively

c.   Electrically nonmetallic rigid connection and piping and electromagnetic nonmetallic stiff connection

d.   Each of the preceding options

17. A busway is not allowed to be erected in any of the subsequent sites unless it is specifically designated as an exception:

a.   Outside

b.   In wet areas

c.   In areas with high humidity

d.   Each of the preceding options

Answer Key of Quiz Number 2

1. Ans: 60 volts

2. Ans: 125 percent

3. Ans: 3 feet

4. Ans: 60 amperes

5. Ans: 150 volts

6. Ans: 24 inches

7. D

8. D

9. A

10. B

11. D

12. A

13. A

14. A

15. D

16. B

17. B

Quiz Number 3

1. Find the ampacity of the conductor provided below data:

Eight (8) modern-carrying conducts are in the raceway, and the outside temperature is 125 degrees Fahrenheit. Conductors were type 500 kcmil copper. Conductive insulating is THWN. The length of the conduit is 50 feet. Third, the installation site is a damp one.

2. Determine the utmost quantity of one-inch AWG XHHW compacted components that are allowed to be housed inside an industrial size 3 inches rigid Grade 40 polyvinyl chloride (PVC) conduit that is longer than 24 inches long.

3. The wiring of the three-way and four-way switching must be done such that the switching may be done either _____ or _____.

4. Class ____ dangerous areas consist of those whereby combustible water-produced vapor or combust water-produced fumes are or might be floating in the air in amounts sufficient to form combustible or ignitible compositions. _____ dangerous spots are situations in which combustion water-produced fumes are or might be available.

5. Where is the maximum protection from overcurrent permitted for the safety of electric heated space technology that uses an electromagnetic resistance type?

6. When secured with a circuit breaker designed for 15 amps, a branch circuit supplied by wires of size 12 AWG is considered to be a branch circuit with a rating of what?

7. GFCI protection is required for every one of the 125-volt, separate-phase, 15- or 20-ampere socket receptacles that are located within at LEAST _____ of the outermost portion of an apartment's bathtubs or showering stall.

8. It is acceptable for computer equipment to be attached to a branching channel by a control-supply cable that is specified, as long as the overall length of the cord does NOT surpass _____.

9. Equipment for infrared warmth used in industrial and commercial environments is required to have an overcurrent security rating of _____?

10. Conductors that are directly buried and emerge beyond level must be covered by passageways or shields that extend up to a minimum of _____ above the final grade.

11. When authorized for at LEAST _____ or more, all three-phase, four-wire, 480Y/277-volt electrical connections are required to provide ground-fault prevention for every connection terminating method. Alternatives to this rule are not considered.

12. According to the National Electrical Code® (NEC®), what is the smallest size wire type apparatus grounded component that must be used for a branch circuit carrying 50 amperes?

13. Metallic conduits and metallic board structures are required to be positioned with a distance of at least _____ between the wall and the tube or panelboard when the apparatus is located in an uncovered position in interior locations where the surfaces are often cleaned, such as laundry facilities and vehicle washes.

14. When making connections to luminaires, the use of adaptable metallic conduit (FMC) with a standard diameter of 3/8 in. is allowed to enclose tap electrics; however, the total length of the FMC must NOT go above which of the following?

15. When a commercial size 1/2 in. solid steel conduit (RMC) is placed in a Class I site, and a 1/2 in. funnel sealing is needed, the minimum thicknesses of the sealing material shouldn't be below _____?

Answer Key of Quiz Number 3

1. Ans: 178.2 amperes

2. Ans: 21

3. Ans: only in the ungrounded circuit conductor

4. Ans: 1

5. Ans: 60 amperes

6. Ans: 15 amperes

7. Ans: 6 feet

8. Ans: 15 feet

9. Ans: 50 amperes

10. Ans: 8 feet

11. Ans: 1,000 amperes

12. Ans: 10 AWG

13. Ans: 1/4 in

14. Ans: 6 feet

15. Ans: 5/8 in

Quiz Number 4

1. One building receives electricity from another structure on a non-industrial property with a single owner. A 100-ampere circuit breaker guards the subsurface feeder in the first building. Because qualified individuals are not always accessible to maintain the installation, the second building's disconnecting mechanism must be one of the following:

    a. Location within the second building, which is not necessarily close to the entrance for conductors, is essential.

    b. Outside or within the building, but preferably at the point where the conductors enter the structure closest to that location.

    c. The initial building's circuit breaker

    d. Situated outside the structure, close to where the conductors enter the structure.

2. Three bathrooms in a single-family home have one receptacle outlet, a fan, and a lighting fixture. The lighting fixture, fan, and outlet are all put on a separate 20-amp circuit in one of the restrooms. Which of the following is the minimal quantity of 20-amp circuits needed for this house to supply the bathrooms?

    a. Two

    b. Three

    c. Four

    d. Five

3. In a 3-wire single-phase system, between the ungrounded conductor and the neutral must be a rated voltage of 120 volts, while that between the two ungrounded conductors must be:

    a. 100 v

    b. 240 v

    c. 300 v

    d. 650 v

4. The middle conductor in a single-phase, having a three-wire system, has to be one of the following:

    a. Hot

    b. Under the ground

    c. Grounded

    d. None

5. Which of the following equations best describes the current flowing through the neutral of a single-phase 3-wire 120/240 volt electrical system?

    a. The current differential between the two ungrounded conductors.

    b. The total current passing across the two unground conductors.

    c. The first ungrounded conductor's current is divided by the second ungrounded conductor's current.

    d. 240 volts divided by 120 volts

6. Which of the following describes the hot conductors in a single-phase, three-wire electrical system:

    a. Neutral conductors

    b. Earthed conductors

    c. Preferred conductors

    d. Conductors ungrounded

7. Which of the following additional installation techniques should be utilized when a rod electrode is needed for grounding reasons, but a layer of rock prevents the rod from being pushed into the ground?

    a. Connect to the closest building's steel structure.

    b. Join the steel water main.

    c. Insert the rod into a trench at least 2 1/2 feet deep.

    d. Cover the rod with at least 18 inches of steel conduit before burying it.

8. In which of the following cases is it necessary to ground the non-current-carrying metal parts of a portable drill with a plug and cord?

    a. The source of electricity is higher than 150 volts to the ground.

    b. The drill is intended for home usage.

    c. The drill is being utilized in a potentially dangerous area.

    d. All of the aforementioned

9. Which of the following is true for the largest overcurrent device size necessary to safeguard a 5-foot length of 1-inch liquid tight flexible metal conduit without equipment grounding conductors and ground listed fitting terminations?

    a. 20 amperes

    b. 40 amperes

    c. 60 amperes

    d. 100 amperes

10. It is not necessary for the bonding jumper between a supplementary electrode and service equipment to be any bigger than:

    a. 2

    b. 10

    c. 9

    d. 6

11. Which of the following outlets is an example that might be attached to a branch circuit for small appliances?

    a. Automatic garage door opener outlet in the garage ceiling

    b. Any container within 20 feet of the kitchen

    c. A dining room electric clock hooked in

    d. A blow dryer electric

12. Which methods should be used to identify terminals attached to a grounded conductor?

    a. Identification must be mostly white.

    b. The connection must use a terminal screw that is green in color and difficult to remove.

    c. Identification must contain a metal tag with an engraving.

    d. None of the above.

13. It follows that there may only be a finite number of switches and circuit breakers in a building if the power is cut off to it:

    a. 6

    b. 9

    c. 11

    d. 21

14. Installing open conductors that aren't service entry cables below one of the following is prohibited:

    a. 4 feet from the level

    b. 6 feet below the level

    c. 9 feet below the level

    d. 10 feet from the level

15. Which of the following describes the total number of underground conductors for an outdoor lighting circuit on a single common neutral conductor?

    a. 4

    b. 7

    c. No limit is indicated

    d. The underground conductors in this scenario are not allowed.

Answer Key of Quiz Number 4

1. B: Section 250. 26 The component intended to be grounded for grounded AC site wiring systems is required to conform to the following specifications:

- A single conductor with a single circuit and two wires
- Single-phase, three-wire, with the neutral component as one of the wires
- Multiphase devices that share a single wire, known as the neutral conductor, among all of the phases
- In multiphase systems, the component of the phase that is grounded is referred to as "that phase."
- Multiphase devices that only use one phase, where the neutral conductor is used

2. B: NEC section 225.32, The disconnecting means must be installed either on the interior or exterior of the building or structure that is being serviced or at the point where the conductors enter or exit the building or structure. The means for disconnecting must be located in an area easily accessible and close to the place where the conductors enter the system.

3. B: The NEC section 250 discusses the general guidelines for grounding and bonding of electrical wiring, as well as the specific requirements listed.

1. Grounding must be necessary, permitted, or not permitted for certain types of systems, circuits, and equipment.
2. In grounded devices, the conductor of the circuit that is to be grounded
3. Where the grounding connections are located
4. Different kinds of grounding and connecting conductors and electrodes, along with their dimensions
5. Various strategies for grounding and connecting
6. There are certain circumstances in which guards, separation, or insulation can be used in place of grounding.

4. C: NEC section 250.26

5. A: NEC section 250

6. D: NEC section 250

7. C: NEC section 250.53 (F), It is required that the ground ring be installed at a depth of at least 750 millimeters (30 inches) underneath the surface of the earth.

8. D: NEC section 250.114/1, Exposed, typically non-current-carrying metal components in cord-and-plug-connected instruments are required to be connected to the equipment grounding conductor in any one of the following conditions: in hazardous (classified) places; in areas where there is a potential for fire or explosion.

9. C: NEC section 250.118 6(C), For measurement designators 21 using 35 (exchange sizes 3/4 employing 11/4), the circuit conducting materials offered in the conduit has been safeguarded by overcurrent machines rated at no more than 60 amps, whereas there is no adaptive metallic conduit, flexible metal-based tubing, or liquid-tight accommodating metal conduit in the true ground-fault present path for measured designators 12 using 16 (trade dimensions 3/8 through 1/2).

10. D: NEC section 250.53 (E), The portion of the bonding connector acting as the sole connection to the supplementary grounding electrode doesn't need to be thicker than a 6- or 4-gauge copper or aluminum wire, respectively, when the supplementary electrode is a plate electrode. This is true regardless of the shape of the extra electrode, which could be a rod, conduit, or plate.

11. C: NEC Section 210.52 B (2): To power additional appliances and illumination on gas-fired intervals, ovens, or light-skinned-mounted preparing food units, a contact instrument must be added at the outlet to accept an attachment plug or to allow direct connection of electrical utilization machinery that is intended to mate with the that corresponds contact device.

12. A: NEC section 200.9, Terminals intended to receive a grounded conductor in devices or utilization equipment with polarized connectors should be colored significantly white or silver. Other terminals' identifying labels must be a distinct color.

13. A: NEC section 230.71 (A), Disconnecting means installed as part of listed apparatus and used exclusively for the following are not a service detaching means for this section:
   1. Instrumentation for monitoring power usage.
   2. Surge-protective gadget(s).
   3. Ground-fault protection system control circuit.
   4. Power-operated method of disabling utility connections.

14. D: NEC section 225.18, When installing open wires or open multiconductor lines not exceeding 1000 volts, nominal, the following clearances must be met:

1. Accessible to foot traffic only and located 3.0 m (10 ft) above completed grade, walkways, or any other surface or projection where physical interaction is possible and the voltage is not greater than 150 volts to earth.

2. If the voltage does not surpass 300 volts to ground, it can be safely installed at the height of 3.7 m (12 feet) above ground in domestic areas, roads, and business areas that do not see heavy vehicle traffic.

3. 3.7 m (12 feet) — for locations in that category where the voltage is greater than 300 volts to earth 4.5 m (15 ft)

4. 5.5 m (18 ft) — over territory used for vehicular traffic such as sidewalks, alleyways, roadways, parking lots with commercial trucks, driveways on non-residential property, and agricultural, grazing, woodland, and orchard land.

5. Distance of 7.5 m (241/2 feet) above train track rails.

15. C: NEC section 225.7 (B): The highest net load current that can be determined across the conductor acting as the neutral and all unsupported conductors linked to any single phase of the circuit must not be less than the magnitude of the neutral conductor.

Quiz Number 5

1. When variations are not appropriate, the overcurrent prevention instrument that protects this circuit must have an output voltage of no less than _____, while a feeder provides a constant load of 240 amperes.

2. The electrical capacity of an electrical conductor is defined as the maximum current, measured in amperes, that a conductor is capable of carrying consistently under certain circumstances of usage without breaking its rating. _____?

3. The inside of any enclosures or raceways that have been constructed underground is to be regarded as a site that is a _____?

4. The disconnect switch enclosure must be designated as suitable for installation in Class II, Division 2 environment. _____?

5. When installing a new power supply in a structure that is not a one- or two-family home, the following requirements must be met: No lower than ONE 125-volt, separate-phase, 15- or 20-ampere-rated receptacles outlets must be situated within a minimum _____ of the power apparatus.

6. The fluid-tight flexibility metallic conduit, also known as LFMC, is required to have a secure connection made to it WITHIN HOW FAR of each cabinet or different conduit expulsion?

7. Situated in one structure or a group of different boundaries, the service disconnection mechanisms for each electrical supply must include but are not limited to not exceeding a certain number of switching or breakers.

8. Find out how much current can be carried by four dimensions 1/0 AWG THW copper power-carrying components when they are placed in a shared raceway at a normal temperature of eighty-six degrees Fahrenheit.

9. The full-load operating current rating for a 208-volt, 3-phase, 50-horsepower, squirrel-cage, continuous-duty, alternating-current (ac) motor is what?

10. You have concluded that a conductor's estimated permitted ampacity permits it to transport a current of 75 amperes securely. What is the MAXIMUM amp grade that the current overload might have? A protective device permitted by the NEC® to safeguard this circuit in situations when the circuit in question is neither a motor circuit nor a component of a multioutlet branches circuit that supplies more than one outlet.

11. What is the UTMOST range that the disconnect switch for a transportable amusement ride may be positioned from the technician's workstation in the attraction's design?

12. Continuous load is a load where the maximum current on the circuit is assumed to continue for ____ hours or more.

13. If the entire length of the pipe constriction is less than 24 inches, you are allowed to fill no more than what percentage of the channel's cross-section region?

14. Determine the smallest trade size of electrically metallurgical tubing (EMT) that will be needed to encapsulate eight (8) 6-gauge AWG conductors made from copper with THHW insulating when placed in an electrical conduit run that is fifty feet long.

15. When a pull box that is going to be placed is going to include conductors size 4 AWG or bigger, and when a straight draw of the participants is going to be made, the overall length of the container shall NOT be shorter than ____times the trade-off dimension of the biggest conduit accessing the box. This is to ensure that the conductors can be pulled properly.

Answer Key of Quiz Number 5

1. Ans: 300 amperes

2. Ans: it's the temperature rating

3. Ans: Dry

4. Ans: Dust tight

5. Ans: 50 feet

6. Ans: 12 inches

7. Ans: Six

8. Ans: 120 amperes

9. Ans: 143 amperes

10. Ans: 80 amperes

11. Ans: 6 feet

12. Ans: Three

13. Ans: 60 percent

14. Ans: 1 1/4 in.

15. Ans: Eight

Quiz Number 6

1. When given a choice between many conductive routes, the electrical current will always choose the one with the least resistance.

   a. True

   b. False

2. Metal components must be grounded to a proper grounding electrode for harmful ground-fault current to be diverted into the earth and away from people in the event of a ground fault, saving them from electric shock.

   a. True

   b. False

3. The grounding conductor for an additional grounding electrode (such as a machine tool's ground rod) must securely carry any fault current that could be applied to it. Depending on the circumstances, this is achieved by sizing the conductor by Table 250.66 or Table 250.122.

   a. True

   b. False

4. Equipment has to be grounded for the circuit protection device to receive enough fault current to open and clear the ground fault fast. For instance, a 120V ground fault to a metal pole connected to a 25-ohm ground rod can trip a 20A circuit breaker and de-energize the system.

   a. True

   b. False

5. Electrical equipment must be grounded to guarantee that a ground fault-induced harmful voltage on metal components can be brought down to a safe level.

   a. True

   b. False

6. Connecting traffic signal pillars and barriers to the implementation covers to a grounding electrode is necessary to reduce potentially dangerous volts on metallic materials induced by a ground fault to an acceptable level.

   a. True

   b. False

7. The safe reduction of a harmful voltage on metal components caused by a ground fault is ensured by grounding metal sewer covers to an appropriate grounding electrode.

    a. True

    b. False

8. In the event of a ground fault, the potentially dangerous volts on metallic components must be eliminated or reduced to safe levels, and this can only be achieved by connecting all service equipment to a grounding electrode.

    a. True

    b. False

9. Service equipment is grounded to an electrode to guarantee that metal components vulnerable to a ground fault maintain the same voltage as the earth.

    a. True

    b. False

10. To keep the electricity flowing through the system, the service devices should always be grounded to an electrode.

    a. True

    b. False

11. By grounding service equipment, any metal components that people may come into touch with are constantly at or close to zero volts concerning the ground (earth).

    a. True

    b. False

12. The purpose of grounding the metal components of independently derived systems is to guarantee that during a ground fault, the voltage, measured between the electrical installation's metal components and the earth, stays at the same potential.

    a. True

    b. False

13. Separately derived systems need to be electrodes to guarantee that a ground fault's potentially deadly voltage on metal components may be eliminated or decreased to a safe level.

    a. True

    b. False

14. The term "ungrounded system" refers to a system that is not grounded, meaning that the independently developed system and its metal container are not connected to the ground (earth).

    a. True

    b. False

15. Inadequately grounding the metallic coating of transformers to the surface of the electrode can result in a potentially dangerous potential difference between the metallic materials of numerous systems that were separately developed. This can happen if the casing of the transformer is metal.

    a. True

    b. False

Answer Key of Quiz Number 6

1. B Note for review: A person touching a grounded but electrically charged metal pole will feel a current run through their body of between 90 and 120 mA, which is more than enough to electrocute them. Remember that current splits and passes via each separate parallel route in parallel circuits.

The only method to ensure that an installation is protected against a ground fault is to attach the electrical equipment to a reliable ground-fault current channel, which will allow the fault current to swiftly release the circuit protection device. NEC 250.2 and 250.4(A)(3)

2. B Review Note: Kirchhoff's current law governs how the current splits and moves via each parallel path in parallel routes. Therefore, if given a choice between many conductive routes, current will follow each one. In a parallel circuit, a lower resistive channel would allow for greater current flow than a higher resistive path, but the topic being asked is different.

3. B Review Note: A ground fault that utilizes the earth as the fault return channel to the power source is incapable of conveying enough current to remove the ground fault [250.4(A)(5)] and would cause unsafe voltage between the metal components and the earth.

4. B Review point: The NEC® [250.54] does not mandate that an additional electrode be sized. The circuit voltage and the resistance of the earth determine how the current flows through the grounding wire into the earth and to the power source during a ground fault. The grounding conductor for a supplemental electrode is not sized in compliance with the NEC® [250.54] because the earth's resistance is too high for it to serve as a reliable ground-fault current channel [250.4(A)(5)].

5. B Review Note: According to [250.2 and 250.4(A)(3)], bonding the traffic signal poles and handhole covers to an efficient ground-fault current channel is the only method to make this installation safe from a ground fault.

6. B Review note: The earth cannot act as a reliable ground-fault current route; therefore, grounding metal components to it will not lower the voltage on metal components caused by a fault [250.5(A)(5)]. For instance, the current going through the grounding conductor into the earth and to the power supply would be just 4.8 amperes, which is insufficient to trip the circuit breaker in a 120-volt circuit with a ground rod resistance of 25 ohms. The only way to ensure this installation is safe from a ground fault is to connect the electrical equipment to a reliable ground-fault current path. In this way, the fault current will be more than enough to open the circuit protection device and clear the ground fault without delay, removing the risk of touch voltage [250.2 and 250.4(A)(3)].

7. B Review point: Based on Section [250.5(A)(5)], you are finding that this test confirms in many ways that grounding metal parts does not reduce the voltage on metal parts that could result from a ground fault. However, you must find that the earth can act as an efficient ground-fault current path.

8. B Review Note: According to [250.2 and 250.4(A)(3)], the only method to allow this installation to be safe from a ground fault is to connect the metal pieces to an efficient ground fault current channel or isolate the manhole cover from any electrified components.

9. B Review Note: System voltage stabilization does not include the earth. The grounding of the utility secondary winding stabilizes the voltage [250.4(A)(1)].

10. B Review remark for Sections 250.2, 250.4(A)(3), and 250.24(C).

11. B.

12. B Review Note: During a ground fault, the earth does not create or maintain a zero-difference of potential between metal components of electrical equipment and the earth.

13. B Review point: According to the NEC®, independently derived systems that are not grounded must have their metal cases grounded to an electrode [250.30(B)(1)].

14. B The best way to make this installation safe from a ground fault is to use a system bonding jumper to connect the metal components of the separately derived system. This will allow the ground fault current to swiftly open the circuit protection device and eliminate the ground fault [250.2, 250.4(A)(3), and 250.4(A)(3)].

15. B Review point: Grounding a transformer's metal casing to a grounding electrode is not required to lower the potential difference between the metal components of several separately generated systems. This is so because the metal components of the individually generated systems are identical. [250.4(A)(3)] All metal components of electrical systems must be bonded to an efficient ground-fault current channel. The metal casing of every individually developed system must be connected to an appropriate grounding electrode, according to the NEC® [250.30(A)(3) and (7)].

Quiz Number 7

1. Situations in which the motor's controller also performs the function of a disconnection imply that the controller is required to disconnect all of the wires.

2. It is essential to have an adequate number of _____ multipurpose containers provided in the guest bedrooms of lodging establishments and sleeping areas of dormitories and that these containers be easily accessible to guests.

3. Class _____ sites are dangerous as a result of an abundance of rapidly ignitible fibers or as a result of the handling of products that create flammable flyings, but in which these kinds of fibers or flyings are not expected to be in dispersion in the atmosphere in numbers adequate for creating ignitable mixes.

4. A specified _____ is required to be placed in or on the switchboards and the panel boards of all rescue circuits.

5. There is a limit to the amount of load that may be put on each line of a single-phase, 240V 15 kVA reserve generation.

6. When it comes to electric space warming equipment, the heating wires must be provided in their whole, including manufacturer-assembled non-heating leads of not less than the specified length.

7. General-purpose the load on re for commercial structures is supposed to be estimated at never below _____ per outlet after consumption variables are put into effect.

8. Compute the maximum quantity of 14-inch AWG THHN, or conductors that may be allowed to be put in a commercial size 3/8 in. flexibility metallic conduit (FMC) provided the FMC includes external connections and the conduit itself already includes a naked dimension 14 AWG a foundation component.

9. When a submerged pump is installed in a good casing made of metallic material, the good casing is required to be _____ the pump loop apparatus grounding circuit.

10. When employed in a dry environment, wires with a certain type of encapsulation have a magnitude greater than when they are utilized in a damp environment.

11. Basement wiring is not allowed to be built beneath a swimming pool that is permanently attached or located inside _____ of the pool, provided the wire is being used to supply an appliance that is linked with the water.

12. A single-phase, 125-v socket installed to service a washing machine that will be located in the laundry facility of an apartment must be placed beyond at least HOW FAR of the planned location of the device?

13. A separate-pole switch and/or a three-way switching are considered equivalent to _____ conductor(s) for the reason of assessing the conductor's capacity in an electronic box. This is determined by taking into account the conductor that is attached to the switches in its greatest capacity.

14. In conditions in which the temperature of the surrounding air does not play a role, the highest permissible magnitude of a single copper the instructor of eight AWG with the protection of type FEPB that is added in the ambient air will be _____?

15. Unless the characteristics of the equipment that the motor propels are enough that it prevents the motor from operating constantly, all motors are to be regarded as being in the category of _____.

Answer Key of Quiz Number 7

1. Ans: Ungrounded

2. Ans: Two

3. Ans: III

4. Ans: surge protective device (SPD)

5. Ans: 62.50 amperes

6. Ans: 7 feet

7. Ans: 180 VA

8. Ans: Four

9. Ans: connected to

10. Ans: THHW

11. Ans: 5 feet

12. Ans: 6 feet

13. Ans: Two

14. Ans: 80 amperes

15. Ans: continuous duty

Quiz Number 8

1. When threading (IMC) in the field, a conventional eliminating die with a decrease of _____ per foot must be employed.

2. Wherever a subjected cable of type NM is routed through the ground, the cable must be protected by being encased in an endorsed intermediary that extends at least _____ beyond the floor.

3. What is the smallest size of copper flooring electrode circuit that must be used when connecting to the concrete-encased reinforcement structure steel whenever an ac power supply is provided with a total of four consecutive pairs of dimensions 500 kcmil aluminum unsupported circuits?

4. If a motor with more than one horsepower has a rise in temperature of 50 degrees Celsius, shown on the nameplate, the overload device must be set to break at NO greater than ____ of the engine's full-load amp capacity to comply with the requirements for choosing the overloading mechanism.

5. What is the longest length of a pliable wire that may be utilized to power the motor of a looping system on a residential pool? What is the longest possible length of the chord?

6. The following must be true of facilities that have electronically heated ceilings: bathrooms, kitchens, and areas with hydromassage bathtubs.

7. What is the estimated voltage drop on the electrical system when an 80-ampere, 240-volt, one-phase demand is situated at a distance of 200 feet from a panelboard and is powered by size 3 AWG wires made of copper having THWN the insulating material (K = 12.9).

8. Plate grounding electrodes must be placed at a depth that is NOT less than (fill in the blank) beneath the outermost layer of the earth.

9. According to the NEC, it is illegal to work at or near a gas station's gasoline pumps. A radius from the motor fuel pumps the container outside and up to an elevation of 18 inches above-level, defining this region.

10. Each intermediary run into or out of Class I, Division 1 protection that included arcing equipment has to have an endorsed seal inside a minimum of _____ of the enclosure.

11. If you're using AWG sizes 6 or bigger for your grounded (neutral) conductors, you can mark their ends during assembly using ____ colored phase tape.

12. Single wires in Class 3 are not allowed to be _____ in diameter.

13. A metal connection container has a capacity of 27 cubic inches and has an overall number of six (6) wires of size 12 AWG; nevertheless, the box does not contain any ground wires, equipment, or fittings of any kind. It is necessary to place more wires in the box that have a gauge of 10 AWG. How many conductors of a size 10 American Wire Gauge (AWG) may be inserted into this device at one time?

14. What is the lowest capacity grounding cable allowed for this kind of setup when connecting the strengthening steel of a pool for swimming?

15. What is the LARGEST possible trade size conduit that can be installed beneath the extension circuit panelboard of an apartment building and underneath a movable home ground to encapsulate a mobile home supplying cord?

Answer Key of Quiz Number 8

1. Ans: 3/4 in.

2. Ans: 6 inches

3. Ans: 3/0 AWG

4. Ans: 115 percent

5. Ans: 3 feet

6. Ans: GFCI protected

7. Ans: 7.84 volts

8. Ans: 30 inches

9. Ans: 20 feet

10. Ans: 18 inches

11. Ans: White

12. Ans: 18 AWG

13. Ans: Five

14. Ans: 8 AWG

15. Ans: 1 1/4 in

Quiz Number 9

1. It is the current, determined by amperes, that an electrical conductor can carry repeatedly during normal conditions without surpassing the highest possible temperature specification of the conductor is referred to:

    a. Maximum Amperage

    b. Flow in a Conductor

    c. Intensity

    d. Convection amplifiers

2. Which of the options that follow best describes a trustworthy conductor used to assure the needed conductance amongst metal elements that are to be electronically connected?

    a. Linker

    b. Assembling Jumper

    c. Electrically insulated wire

    d. Connector for Branches

3. Which categories best describe provisions of the Code that detail particular behaviors mandated or forbidden?

    a. Norms that provide leeway

    b. Inflexible regulations

    c. Procedures for Inspection of the Setup

    d. Use Authorization

4. Which of the items listed below may be avoided by reducing the overall number of circuits included inside a specific enclosure?

    a. The results of a short circuit in a single circuit are reduced to a minimum.

    b. Prevents further over-expansion

    c. Facilitates access to skilled labor

    d. It sets a benchmark for the AHH to follow

5. Which of the items that follow is an example of a device that can de-energize a circuit or a section of circuitry within the period that has been predetermined when the electrical current ground voltage surpasses the values that have been judged to be appropriate for a Class A device?

    a. Breaker or Circuit

    b. Fuse

   c. Interrupter for ground faults

   d. Switch for adjusting the voltage

6. Which of the subsequent terms best describes a configuration consisting of two or more unsupported wires with equivalent energy flowing across them along with a neutral conductor?

   a. Regulating system

   b. As a Feeder

   c. Electricity in a branch

   d. Feedback loop

7. Which of the options best describes an assembly consisting of at least two single-pole fuses?

   a. Several fuses

   b. Connector with multiple taps

   c. Switching mechanism

   d. Printed circuit board

8. Which of the choices that follow best describes a component of a motor that prevents it from overheating while still being an essential part of the motor itself?

   a. In-line fuse

   b. Protector against ground faults

   c. Heat shielding

   d. Indicator of Shunt Trip

9. Which of the phrases that follow is used to describe a branch circuit that consists of more than one unsupported conductor, each of which has a possible distinction between it and the other unsupported conductors, and a grounding conductor that has an equal prospective imbalance between it and each of the unsupported conductors?

   a. System of continuous loop feeding

   b. Multiple wires

   c. All-purpose circuit

   d. A branch circuit that is regulated

10. A load is considered to be continuous if it is anticipated that it will be subjected to one of the four categories of current for at least three hours without interruption:

   a. Constant

   b. Normal

   c. Maximum

d. Restricted

11. Which of the following is the name of the connection device that may be inserted at an outlet to accommodate two or more connection instruments at the same yoke?

    a. Duplex plug

    b. Multiple receptacles

    c. Slice connection

    d. None of the aforementioned

12. Which of these statements is the most accurate definition of circuit voltage?

    a. The average potential difference across two conductors

    b. The greatest possible distinction between two conductors

    c. The equivalent potential difference between two conductors

    d. Total amperes generated across two conductors

13. Which of the items listed below is an electrical switch that may be used to disconnect a circuit or piece of equipment from a preexisting power supply?

    a. Disconnect Button

    b. Circuit Blocker

    c. Cutout

    d. Interrupter toggle

14. Which of the following best describes a big single-panel installation of panels, including mounted toggles, overcurrent and safeguarding devices, and buses?

    a. Panelboard

    b. Switchboard

    c. Switch for the automatic transmission of power

    d. Service panel

15. Which, if any, of the following types of electrical installations are exempt from the requirements of the Electrical Code?

    a. Power distribution wiring for huge industrial machinery

    b. Computer cabling in a commercial building

    c. Telephone cable installed in a flexible conduit

    d. All of the preceding

16. Which of the aforementioned terms best describes a space to which any number of air passageways are attached to form the component of an air-distributing system?

   a. Air-flow box
   b. Ventilation duct
   c. Circulation cavity
   d. Plenum chamber

17. Which of the items listed below must meet the minimal standards for fire resistance on the level of a vault designed to store electrical devices if there is either a space or further floors below it?

   a. 1 hr.
   b. 3 hrs.
   c. 6 hrs.
   d. 12 hrs.

18. Choose the option below that does not fall within the category of a permitted method for attaching electrical components to a brick wall:

   a. Inserting screws into wooden screws in the wall
   b. Employing screws that are reinforced on the back side by plates of metal
   c. Employing molly bolts via thoroughly drilling openings in the wall
   d. Installing lag bolts into lead brickwork anchoring

19. The equipment workspace has to have enough room for a _____ aperture for any gates or hinges screens on the equipment.

   a. 180-degree
   b. 45-degree 60-degree
   c. 90-degree
   d. 60-degree

20. Only conductors classified for each of the following capacities can be utilized in circuits with 100 amps or smaller or with conductors ranging from 2.0 square millimeters to 50 square millimeters.

   a. 131 degrees F
   b. 140 degrees F
   c. 167 degrees F
   d. 176 degrees F

Answer Key of Quiz Number 9

1. A

2. C

3. B

4. A

5. C

6. C

7. A

8. C

9. B

10. C

11. B

12. C

13. D

14. B

15. D

16. D

17. B

18. A

19. C

20. B

Quiz Number 10

1. Which equipment must ground conductors be present when conductors are run in parallel raceways?

    a. Run in separate raceways

    b. Held up every six inches

    c. Run concurrently in each racetrack

    d. Shielded from high temperatures

2. If a solitary 30 amp branching circuit supplies power to a single outlet for non-motorized gear, that receptacle must have a current of which of the following?

    a. 10

    b. 30

    c. 50

    d. None of the above

3. When the building has multiple nominal voltage systems, unsubstantiated feeder conductors must be:

    a. Using a transfer switch to connect

    b. Individually identifiable

    c. Run in separate ducts

    d. Marked with the same color

4. Any surge arrestor with a nominal voltage of fewer than 1000 volts must have a ground connecting conductor that is not less than which of the following sizes:

    a. #9

    b. #11

    c. #12

    d. #14

5. The exterior dimensions of a home, including any porches or garages, are utilized to calculate the area needed to support the lighting load.

    a. True

    b. False

6. Which of the following is true for single-point grounding at the source of a separately derived system?

    a. Each structure and enclosure has its equipment grounding conductor.

b. Aside from one spot, the neutral is shielded and separated from the earth.

c. A grounding conductor for the equipment is run together with the phase conductors.

d. All of the aforementioned

7. The projected load on a feeder, after derating, must be no more than 75% of the total load on the distribution lines.

a. True

b. False

8. A branched circuit's constant load must not surpass 80% of the cable capacitance value without using derating factors.

a. True

b. False

9. The size of the residual current device determines the branch circuit rating.

a. True

b. False

10. A feeder neutral may, under some circumstances, be smaller than the unbiased conductors in a wiring system.

a. True

b. False

Answer Key of Quiz Number 10

1. C: NEC section 250.122 (F),

1. Single Raceway, Auxiliary Gutter, or Cable Trough are the three options for this. If multiple circuit cables are linked together in parallel in an identical raceway, supplemental gutter, or cable tray, then only a single wire-type conduit may be used as the grounding conductor for the apparatus. The amount of the grounding conductor for wire-type equipment must be determined depending on the overcurrent protective mechanism for the feeder or offshoot circuit.

2. Multiple Tracks for Racing. If conductors are installed in numerous raceways and are linked together in parallel, a wire-type apparatus grounding conductor, if used, must be implemented in each speedway and must be connected in parallel. This applies even if the conductors are installed in separate raceways. The amount of the equipment grounding conductor installed in each raceway must be determined following the specification of the overcurrent protective device used for the feeder or branch circuit.

2. B: NEC section 210.21 (B)

3. B: NEC section 215.12 states that every single unfounded conductor of a feeder must be designated by phase or line and scheme at all dismissal connection and splice locations in a premises electrical system that is provided by more than one nominal voltage system. This is required at any place where there is a dismissal connection or splice.

4. D: NEC section 280.21

5. B: NEC section 220.112: The minimal lighting load is to be calculated using the determined floor area and a unit load not less than that prescribed for non-residential occupancies. Motors with a horsepower rating of less than 1/8 horsepower that are wired into a building's lighting system are classified as general lighting loads.

6. D: NEC section 250.184 (B), The following shall be observed when employing a grounded neutral system with a singular source of grounding:

1. Power for a grounded neutral device with a singular point of connection may come from either option (a) or (b):

   • Independently determined methodology

- An alternative to traditional single-point grounded neutral systems is a multi-grounded neutral system in which an equipment grounding wire is connected to the multi-grounded neutral conductor at the place of origin.

2. The device requires a grounding wire to function properly.

3. The ground electrode must be connected to the system neutral via a grounding wire cable.

4. It is good practice that the electrode conductor should be connected to the grounding conductor of the device via a bonding connector.

5. Every building, structure, and apparatus enclosure needs to have a grounding wire installed.

6. Where loads are provided from the neutral to the phase, only then will a neutral conductor be needed.

7. Wherever it is installed, the neutral wire must be completely separated from the ground.

8. The phase wires must include a grounding conductor for the apparatus that meets the requirements in (a), (b), and (c) below.

   - Shall not bear a constant weight
   - Possibly protected, possibly not
   - Be equipped with adequate ampacity for problem current responsibility.

7. B: NEC section 220.40: After applying any demand factors allowed, the determined load of a feeder or carrier shall not be less than the total of the loads on the branch circuits.

8. B: NEC section 210.19 A (1): A branch circuit's minimal conductor capacity must have an ampacity equal to the sum of the noncontinuous load and 125% of the continuous load whenever it serves either or both of these load types.

9. A: NEC section 210.3

10. A: NEC Section 220.61: It is acceptable to employ only the biggest load(s) that will be used at one time when determining the overall load of a feeder or service when it is doubtful that two or more noncoincident loads will be in use concurrently. If a motor is included in the noncoincident load but isn't the biggest of the weights included in the noncoincident load, 125 percent of the motor's load must be used in the computation.

Quiz Number 11

1. It is required that any raceway or cable holes in boxes or conduits that are not in use be sealed off, and the protection that results must be one of the following:

    a. Offering a level of protection at least comparable to that offered by the wall of the box or conduit

    b. At a depth that is equivalent to that of the wall framing

    c. More comprehensive than the protection offered by the box or conduit on their own

    d. Capable of serving as an effective fire barrier

2. To install high-voltage conductors in tunnels, you need either an iron conduit or metallic passageways. In addition, you need which of the items listed below:

    a. EMT cable

    b. Copper–clad aluminum conductors

    c. Aluminum conductors

    d. Type MC cable

3. The following are some of the labels that are commonly found on equipment and terminations:

    a. The installer's initials

    b. Tag de service

    c. Torque-increasing

    d. Terminal identifications

4. Which of the following should be considered the minimum acceptable fire rating for walls, floors, and doors, including equipment with a nominal voltage of more than 600 volts?

    a. 1 hr.

    b. 2 hrs.

    c. 3 hrs.

    d. 6 hrs.

5. Which of the items listed below must splice to ensure that they are electrically stable before soldering?

    a. Sanded

    b. Mechanics–based joint

    c. No sharp corners

    d. Flux-coated

6. The airflow in a ventilation system should be controlled using electrical controls placed in one of the methods that follow:

    a. With an external vent

    b. The opposite

    c. Demand-based rationing

    d. Stopped

7. Which of the subsequent descriptions best describes walls made of concrete and brick?

    a. Grounded

    b. Insulators

    c. Wet locations

    d. Dry locations

8. The live portions of electrical equipment that operate at a voltage of ___ volts or more are required to be kept under wraps unless otherwise indicated.

    a. 150

    b. 100

    c. 50

    d. 200

9. Which of the aforementioned colors should a high-let conductor in a three-phase, four-wire delta auxiliary be?

    a. White

    b. Black

    c. Green

    d. Orange

10. If a given article or section does not specify a conductor material, the material in question is presumed to be one of the following options:

    a. Copper-clad aluminum copper

    b. Aluminum

    c. Copper

    d. No assumptions made

11. Which of the items on this list is not an example of a standard value for an excessive current device?

    a. 150 amp.

    b. 110 amp.

    c. 175 amp

    d. 1500 amp.

12. Which of the following options is not a listing format for conductor sizes?

    a. Circular millimeters

    b. Size or thickness

    c. AWG (American Wire Gauge) or millimeters

    d. AWG (American Wire Gauge) or round mils

13. Which of the subsequent applications is acceptable for the use of type AC cables?

    a. In work that is exposed to the elements

    b. In the case of covert labor

    c. Either A or B

    d. None of the preceding options

14. Conservatories around electrically powered conservatories must be controlled by one of the following to be considered accessible:

    a. Locks that require a key

    b. Locks that require a combination

    c. Locks with keypads that need a properly authorized individual to have the code to unlock the enclosure.

    d. Any of the preceding options

Answer Key of Quiz Number 11

1. A

2. D

3. C

4. C

5. B

6. B

7. A

8. C

9. D

10. C

11. B

12. D

13. C

14. D

Quiz Number 12

1. If a neutral bus is necessary, it must be of sufficient size to carry all of the neutrality load currents and must adhere to one of the following requirements:

    a. Have a size that is suitable for transporting harmonic currents

    b. Possess a suitable momentary rating that is compatible with any criteria placed on the system.

    c. Possess a short-circuit rating that is compatible with the requirements of the system.

    d. Each of the preceding options

2. To remove accumulated moisture from low areas throughout the length of a busway run, which of the methods outlined below should be installed:

    a. Drainage plugs and stoppers

    b. Filter drains

    c. One of the options presented above

    d. It is against the rules for Neither-Busway runs to have low spots.

3. Which of these options can be utilized to either raise or decrease the amount of alternating current?

    a. Switch to avoid the bypass

    b. A transformer

    c. Conductor for grounding the system

    d. None of the preceding options

4. Which of the following best describes an appliance of protection used to reduce spike voltages by discharge currents that surge and inhibit ongoing movement of lead current while yet being capable of duplicating these functions?

    a. Surge arrestor

    b. Fuse

    c. Disconnector

    d. Circuit breaker

5. When used in conjunction with polyphase induction engines, heating overload relays serve primarily the aim of protecting from which of the adhering to?

    a. Fire

    b. Electrical interruptions between the phases

    c. Low voltage

    d. Continuously excessive load

6. Which of the categories carries any number of layers that include faux-conducting substances that do not constitute insulation?

    a. Covered

    b. Packaged

    c. Rubberized

    d. Each of the preceding

7. The earthing termination of a grounded connector is required to be connected to a grounded container using which of the following:

    a. A conduit or conduit box

    b. A primary bonding jumper

    c. A neutral substance conductor

    d. A device for bonding jumper

8. It is permissible for rigid non-metallic conduits to have an aggregate amount of quarter bending in a single run that is not greater than which of the following is applicable:

    a. 2

    b. 5

    c. 6

    d. 1

9. In situations in which the surrounding temperature is higher than the one afterward, the conductor's magnitude must be reduced:

    a. 30 degrees C

    b. 38 degrees C

    c. 16 degrees C

    d. 12 degrees C

10. After a capacitor has been separated from its power supply, the capacitor's leftover voltage must be lowered to which of the subsequent voltage ranges within a minute of the capacitance's disconnection:

    a. 50 volts or less

    b. 80 volts or less

    c. 1o V or less

    d. Zero V

11. Which one of the following does not belong to the category of methods for attaching apparatus to concrete?

    a. Shield made of lead

    b. A bushing made of steel

    c. Bolts used for expansion

    d. Each of the preceding options

12. Which of the various switches do you need to have at each place to have independent control over the lights from two separate locations?

    a. Throws with a single pole but two attempts

    b. A single throw with a three-way split

    c. Dual pole, dual throw

    d. A single throw with a double pole

13. Which of the following scenarios is the most prevalent application for mica in the development of electrical systems?

    a. In the control panels of the switchboard

    b. Separators for the communicator bus

    c. Insulators that conduct electricity

    d. None of the preceding options

14. Which of the items below will have the same level of energy consumption as a resistor with 14.4 ohms if you use a bulb that is 1000 watts and 120 volts?

    a. 1100 watts

    b. 140 volts

    c. 104 volts

    d. 120 volts

15. It is possible to guarantee that there is continuity in electricity among conductors by eliminating which of the following elements from the pipe threads:

    a. Enamel

    b. Copper ends

    c. Coating made of rubber

    d. Each of the preceding options

Answer Key of Quiz Number 12

1. B

2. A

3. B

4. A

5. D

6. A

7. D

8. D

9. A

10. A

11. B

12. A

13. B

14. D

15. A

Quiz Number 13

1. PV source circuits inside the PV array may employ single-conductor cable such as _____ and single-conductor cable classified and designated as solar photovoltaic (PV) wire.

    a.   UF

    b.   THHN

    c.   USE-2

    d.   PVC-2

2. The manual switch for the emergency lighting systems is often allowed to be placed anywhere in a movie theatre.

    a.   In the film projection booth

    b.   In a location in the lobby that is accessible

    c.   On the platform or stage

    d.   In any of the following locations

3. If ground-fault protection is not present, it is allowed to ground the meter enclosure to the _____ conductor for a building or structure when the meter enclosure is close to the service disconnecting mechanism.

    a.   Grounded

    b.   Grounding

    c.   Bonding

    d.   Phase

4. All single-phase, 125-volt, 15- and 20-ampere outlets placed within six (6) feet of the top inside edge of the sink bowl in _____ must be protected by a GFCI in residential units.

    a.   laundry rooms

    b.   bathrooms

    c.   garages

    d.   all of these areas

5. The maximum allowed cord-and-plug-attached load to the receptacle is _____ when a 125-volt, 15-ampere receptacle is linked to a 120-volt branch circuit serving two (2) or more receptacles or outlets.

    a.   12 amperes

    b.   15 amperes

    c.   10 amperes

d. 7.5 amperes

6. A 125-volt, 15- or 20-ampere receptacle for under-sink garbage disposal installed beneath the counter in a housing unit must have:

    a. AFCI

    b. GFCI

    c. Protection AFCI and GFCI both

    d. Protection neither AFCI nor GFCI

7. The cable _____ while installing non-metallic sheathed cable encased in a subsurface PVC conduit.

    a. must be Type NMC

    b. must be Type NMS

    c. must be Type NMW

    d. is prohibited

8. What is the maximum allowable length of the 200-ampere busway when overcurrent protection is not given only in a commercial and industrial setting where a 200-ampere busway is connected to a 600-ampere busway?

    a. 100 feet

    b. 50 feet

    c. 75 feet

    d. 125 feet

9. Four (4) size 2 AWG THWN insulation protecting copper wires that transfer electricity with a total length of six (6) feet are included in a service-entrance conduit for a commercial building. Each of these conductors has a permitted ampacity of:

    a. 92 amperes

    b. 100 amperes

    c. 115 amperes

    d. 110 amperes

10. The NEC states that fixed outside snow melting and deicing equipment should be regarded as a _____ load.

    a. Non-continuous

    b. Intermittent

    c. Continuous

    d. Simultaneous

Answer Key of Quiz Number 13

1. Answer: c

The wiring techniques designated for use on solar photovoltaic (PV) system arrays are covered in Part IV of Article 690. Single and open conductors, such as type USE-2 and single conductor cable labeled and recognized as PV wire, can be used in PV source circuits inside the PV array, according to Section 690.31(C)(1). These conductors have undergone evaluation and are approved for usage in rainy and sunny environments.

2. Answer: b

Buildings used as theatres or for similar purposes are subject to the criteria of Article 520. Part V of Article 700's regulations, which apply to controlling emergency lighting circuits, are mandated by Section 520.8 for theatres' emergency systems. A manual control switch for emergency lights must be positioned in the lobby or another easily accessible area rather than in a motion picture projection booth, on a stage, or anywhere else prohibited under Section 700.21.

In the case of a fire or other emergency, or if the usual power supply fails, emergency systems are designed to maintain a particular illumination level for exit routes from the facility.

3. Answer: a

The main specifications for equipment grounding techniques are included in Part VII of Article 250.

By Section 250.142(B) Exception No.2, the meter enclosure may be grounded by connection on the load side of the service disconnect if the following conditions are met: (1) there is no ground-fault protection installed; (2) the meter enclosure is located close to the service disconnecting means; and (3) the grounded service conductor does not have a smaller diameter.

4. Answer: d

According to Section 210.8(A)(7), GFCI protection must be installed to avoid a ground-fault shock hazard regardless of the location or room in the house where the 15- or 20-ampere, 125-volt receptacles are put within six (6) feet of the top inside edge of the sink basin.

5. Answer: a

Branch circuits are generally governed under Article 210. Section II of Article 210 addresses branch circuit and receptacle ratings. A receptacle connected to a branch circuit that supplies multiple receptacles or other outlets must not provide a total cord-and-plug-attached load that exceeds the limit allowed by Table 21(B)(2) to avoid overheating the outlets. According to the chart, a 15-ampere rated receptacle may handle a maximum cord-and-plug coupled load of 12 amperes under these circumstances.

6. Answer: c

Every 125-volt, 15- or 20-ampere receptacle placed within six (6) feet of a sink must be GFCI protected, according to Section 210.8(A) (7), since the presence of water and grounded surfaces increases the risk of an electric shock. According to Section 210.12(A), all 120-volt, 15- or 20-ampere branch circuits supplying outlets in homes' kitchens must have AFCI protection.

The purpose of AFCI devices is to assess resistance to unintended tripping caused by arcing that develops in usage equipment during typical operating settings or by a loading scenario that closely resembles an arcing failure.

7. Answer: d

The inside of all raceways and enclosures placed underground is regarded as a wet site, as defined in Section 330.5. In underground installations, conductors and cables must be of a kind that is approved for use in moist environments.

Part II of Article 334 covers use restrictions for non-metallic encased cables. Type NMC is allowed in dry, moist, and damp environments, according to Section 334.10(B)(1); however, wet sites are not included in this list. Types of NM and NMS cables cannot be put in moist or damp places, according to Section 334.12(B)(4).

8. Answer: b

The installation of busways is permissible and is governed by Part II of Article 368. Suppose the busway with the smaller ampacity is at most fifty (50) feet long. It has an ampacity of at least one-third higher than the larger busway's overcurrent protection rating. In that case, overcurrent protection is not necessary when the busway size is changed for busways only in industrial establishments.

- 200 amps are equal to one-third of 600 amps.
- Observe exemption in Section 368.17(B)

9. Answer: a

Locate the conductor type on the left side of Figure 310.15(B)(16), then look beneath the copper THWN section with a 75°C temperature rating and 115 amperes. The table assumes a raceway with three current-carrying components and an 86°C ambient temperature.

This raceway has four current-carrying conductors; thus, use the appropriate correction factor (derating value) from Table 310.15(B)(3)(a):

- The Ampacity of 2 AWG THWN = 115 amperes (before derating)
- 80 percent of 115 amperes is 92 amperes.

10. Answer: c

According to Article 100, a continual load is expected to endure for at least three (3) hours. Permanently installed outside electro-deicing and snow-melting technology must adhere to the standards outlined in Article 426. Electric snowplows and ice melters permanently installed outdoors are examples of continuous loads for Section 426.4.

When sizing conductors and overcurrent protection for fixed outdoor equipment, it is important to remember that the equipment may run continuously for more than three hours, making it a continuous load.

Quiz Number 14

1. Electrical metallic tubing is not permitted to be used as a grounding conductor for equipment connected to a circuit.

    a. True

    b. False

2. A continuous white outside finish is the only way to identify an insulated grounded conductor size #6 AGW or less.

    a. True

    b. False

3. In which of the subsequent circumstances is the usage of cable of type UF required?

    a. Concrete enclosed

    b. In-ground direct burial

    c. Service acceptance cable

    d. Neither of the preceding options

4. In which of the following situations does it make sense to attach the antenna of a satellite system to a grounding point?

    a. The inside metal water pipe system, located within five feet of the place where it enters the building

    b. The grounding electrode system applicable to the building or structure

    c. A method that can be reached from the outside of the structure

    d. Each of the preceding options

5. The overall connected demand in watts comprised of three electric heaters, two of which are rated at 1000 watts and one at 1100 watts while operating at 120 volts, and the third operating at 800 watts when operating at 240 volts, is as follows:

    a. 2417 watts

    b. 1900 watts

    c. 1160 watts

    d. 4300 watts

6. The mechanism for disconnecting service must make it abundantly obvious which of the following is applicable:

    a. The lowest and highest acceptable voltage ratings

    b. Whether it is open or closed in its current state.

   c. All of the preceding together

   d. Neither of the preceding options

7. In the absence of a clear and conspicuous indication to the contrary, which of the following sites does not allow the use of non-metallic wireways?

   a. In areas of the wire way that would be subject to direct sunlight

   b. In areas where corrosive vapors are present

   c. In areas with high humidity

   d. Each of the preceding options

8. A load is considered to be continuous if it is anticipated that the maximum current will remain constant for a minimum of one of the reasons that follow:

   a. 24 hours

   b. more than 12 hours

   c. more than 6 hours

   d. more than or equal to 3 hours

9. A fuse with a size of which of the following must be used to safeguard a conductor with a rating of 56 amperes

   a. 60 amp

   b. b.33 amp

   c. 28 amp

   d. 44 amp

10. Which of the items listed below can be supplied by feeders as long as they have a common neutral?

   a. One, two, or three sets of feeders that have three wires each

   b. Two pairs of feeders with four wires each

   c. Two pairs of feeders with five wires each

   d. Each of the preceding options

11. Which of the following equations comes closest to describing the minimum component amperes for twin hot, unsupported feeder network conductors flowing to a corporate structure that has an average illumination load of 50,000 volt-amps and is powered by just one phase, 240vV circuit?

   a. 118 amps

   b. 207.39 amps

c. 128 amps

d. 480 amps

12. Which of the following substances has the highest maximum permissible ampacity for nine 8.0 sq mm THHN conductors of copper placed in a conduit and operating at 30 degrees Celsius?

a. 26 amperes each

b. 38.5 amperes each

c. 45 amperes each

d. 77 amperes each

13. Which of the subsequent plugs and sockets can be attached to the circuits for small appliances:

a. A socket in the ceiling of the garage that can accommodate an electric garage door opener

b. A space in the living room that can accommodate a television set

c. A place in the kitchen where an electric clock may be plugged in

d. None of the preceding options

14. Which of the options must be used to safeguard a multi-wire branching circuit that feeds several devices on the same yoke in a housing unit?

a. One fuse

b. Two circuit breakers with a single pole are connected

c. Two circuit breakers with a single pole that are capable of working independently

d. Single single-pole non-fused disconnection

Answer Key of Quiz Number 14

1. B

2. A

3. B

4. D

5. A

6. B

7. A

8. D

9. A

10. D

11. D

12. B

13. C

14. B

Quiz Number 15

1. Branch circuits cannot be created with autotransformers unless the following conditions are met:

   a. Unauthorized people need help to reach the grounded conductor.

   b. The autotransformer runs at a nominal voltage of fewer than 600 volts.

   c. The autotransformer supply system's grounded conductor is electrically linked to the grounded conductor in the supplied circuit.

   d. All of the above.

2. Which of the following is an acceptable identification for an insulated grounded conductor that is size #6 AGW or smaller:

   a. A consistent white exterior finish

   b. A green insulated conductor with three yellow stripes running the length of it

   c. A grey insulated conductor with a single blue stripe

   d. All of the above

3. Which of the following describes the minimum size of a conductor used to ground equipment with an automated overcurrent device rated at 20 amps in the circuit in front of the piece of equipment?

   a. 3.5 sq mm copper

   b. #12 AWG copper

   c. #10 AWG copper

   d. #8 AWG copper

4. An AC circuit with two ungrounded wires may be tapped from a circuit having which of the following characteristics:

   a. A pair of switchgear

   b. A multi-pole automated switch

   c. A grounded neutral conductor

   d. None of the preceding

5. Which of the following requirements must be met for a branch circuit size for an electric 9 kW range?

   a. 15A

   b. 20A

   c. 30A

    d. 40A

6. Which of the following conditions must be satisfied by grounded electrical systems?

    a. Have a connection to the ground that reduces the voltage brought on by line surges.

    b. Have a connection to the ground that will control the voltage during normal use.

    c. The two options above.

    d. One of the options above.

7. Which of the following situations prohibits the installation of a TVSS device?

    a. On a grounded impedance system, first.

    b. When the TVSS falls below the highest continuous phase-to-phase power frequency possible at the application site.

    c. On circuits that have more than 600 volts.

    d. All of the above.

8. The TVSS must be linked in which of the following ways in an independently developed system:

    a. Each ungrounded connection in clause A

    b. Outside, in a conveniently located area

    c. To the grounded delta in the corner

    d. On the first overcurrent device's load side

9. Which of the scenarios qualifies for the installation of a surge arrestor?

    a. On a grounded impedance system, first.

    b. With an electrode on a service with less than 1000 volts.

    c. Either of the options above.

    d. None of the preceding.

10. For proper operation, a 125V solitary countertop receptacle must meet which of the following criteria?

    a. ground-fault interruption

    b. maximum of 20 amps

    c. minimum of 15 amps

    d. All of the above

11. Which of the following describes the minimum size of a conductor used to ground equipment with an automated overcurrent device rated at 40 amps in the circuit in front of the piece of equipment?

    a. #8 AWG copper

    b. #8 AWG aluminum

    c. #10 AWG aluminum

    d. #5.5 AWG cooper

12. Which of the following is approved for supply via multi-wire branch circuits?

    a. Only line-to-line neutral loads

    b. Only one piece of utilization equipment

    c. Ground detectors are installed

    d. None of the above

13. The total cord-and-plug load for a 30-amp receptacle connected to a 30-amp branch circuit feeding two or more outlets shall not exceed:

    a. 24 amps

    b. 16 amps

    c. 15 amps

    d. 12 amps

14. Which of the following is not necessarily about the size of the solitary connection of a grounding electrode conductor linked to a concrete-encased electrode?

    a. Larger than 2.0 sq mm copper.

    b. Attached to an electrode that is at least 2 inches deep in concrete.

    c. Both of the above.

    d. None of the above.

15. Which of the following is allowed for an equipment grounding wire that is connected to circuit conductors?

    a. Copper busbar

    b. Flexible metal conduit

    c. Liquid tight conduit

    d. All of the above

Answer Key of Quiz Number 15

1. C

2. A

3. A

4. A

5. D

6. C

7. D

8. D

9. B

10. D

11. D

12. C

13. A

14. A

15. A

Quiz Number 16

1. Which circuit breakers mentioned below do not have a conventional ampere rating?

    a. 75 amperes

    b. 90 amperes

    c. 110 amperes

    d. 225 amperes

2. In a multi-family residence, a 150 kVA, single-phase transformer with a secondary voltage of 120/240 is installed. The secondary's full-load current rating, in amps, is. _____.

    a. 625 amperes

    b. 265 amperes

    c. 526 amperes

    d. 1250 amperes

3. In general, conductors for overhead service drops must NOT be smaller than _____.

    a. 2 AWG copper

    b. 4 AWG copper

    c. 6 AWG copper

    d. 8 AWG copper

4. To splice conductors or make connections to luminaires or other devices, the MINIMUM length of free conductors that must be left at each junction box is. _____

    a. 4 inches

    b. 10 inches

    c. 8 inches

    d. 6 inches

5. Why is a metallic conduit necessary to provide electrical continuity?

    a. To reduce voltage drop.

    b. To limit galvanic corrosion.

    c. To establish an optimal grounding path.

    d. To reduce electrolysis.

6. The NEC mandates that a galvanized eye bolt be placed NOT less than ____ above the completed grade to serve as the connection point for electrical service.

    a. 8 ft.

    b. 10 ft.

    c. 12 ft.

    d. 15 ft.

7. The ampacity factors for three (3) current-carrying conductors in the raceway do not need to be used when conductors are put in tubing nipple of ____ or less in length

    a. 24 inches

    b. 30 inches

    c. 36 inches

    d. 48 inches

8. Receptacle outlets for countertop surfaces must be placed above residential dwellings, but at most ____ above the countertop or work surface.

    a. 12 inches

    b. 18 inches

    c. 20 inches

    d. 24 inches

9. Which of the following cannot be placed over a stairway's steps?

    a. general-use receptacles

    b. overcurrent devices

    c. switches controlling luminaires

    d. junction boxes

10. By the NEC, whether sockets, appliances, or lighting fixtures are powered from the branch circuit, the rating of a 120-volt, cord-and-plug-connected room air conditioner should not be more than ____ of the rating of the branch circuit

    a. 40%

    b. 50%

    c. 70%

    d. 80%

11. Metal water pipes must be bonded to ____ before being put in or connected to a building or structure.

    a. service device enclosure

    b. the grounded conductor at the service

    c. the grounding electrode conductor

    d. any of the above

12. A minimum of one (1) single-phase, 125-volt, 15- or 20-ampere receptacle must be placed at least ____ from and not more than ____ from the inside wall of any outdoor swimming pool that is permanently erected on residential or commercial.

    a. 6 feet – 20 feet

    b. 10 feet – 20 feet

    c. 6 feet, 6 inches – 10 feet

    d. 5 feet – 20 feet

13. When armored cable, Type AC, is put in and run to sides of rafters of an accessible attic, _____.

    a. the cable-protected boards

    b. the cable must be protected by protective strips, which must be equal in height

    c. either A or B

    d. It is not necessary to have safety railings or running board

Answer Key of Quiz Number 16

1. Answer: a

The typical ampere ratings for fuses and circuit breakers are shown in Table 240.6(A). It is not a typical ampere rating for a circuit breaker to have a 75-ampere rating.

2. Answer: a

Use the single-phase current calculation as indicated to get the single-phase transformer's full-load current rating:

I = 15,000/240 = 625 amperes (150 kVA times 1000)

3. Answer: d

Overhead service-drop conductors must comply with Section 230.23(B) and be no smaller than 8 AWG copper, 6 AWG aluminum, or copper-clad aluminum. Refrain from mistake service-entrance conductors, the conductors that run from the service point owned by the utility company to the account disconnecting method, along with service-drop conductors, which are the overhead conductors that connect the utility electricity distribution network and the service point.

4. Answer: d

According to Section 300.14, each junction box must have at least 6 inches of open conductors available for splices or connecting luminaires or other devices. This guideline aims to provide sufficient conductor slack to enable straightforward termination connections or splices.

5. Answer: c

Metallic conduits must have electrical continuity to provide a reliable ground-fault channel and make the overcurrent protection device work more efficiently. Think about 250.4(A)(4) and (5).

6. Answer: b

The purpose of Section 230.26 is to ensure that conductors have appropriate vertical clearance, avoid physical damage to the conductors, and shield people from unintentional contact with the conductors that might result in shock or electrocution. The service-drop conductors' connection point must always be 10 feet above the completed grade.

7. Answer: a

According to Section 310.15(B)(3)(a)(2), adjustment factors for raceways with more than three (3) current-carrying conductors do not apply when the racetrack is less than 24 inches. When it is

178

more than that, three current-carrying wires are placed in short lengths of channel or tubing. The allowable measure of the concentration of the conductors is quite high since the additional conductors have a minor impact on the heating of the shorter lengths of conduit or tubing.

8. Answer: c

Receptacle outlets installed for countertops or work surfaces in dwelling units may be placed on or above, but at most 20 inches above the countertop or work surface, by Section 210.52(C)(5). This regulation's purpose is to prohibit unattractive extension cables on the countertop or work surface to supply cord-and-plug linked equipment (which decreases the amount of available countertop space). Moreover, outlets placed more than 20 inches above counters or work surfaces may obstruct the installation of the cabinets.

9. Answer: b

The authorized site criteria for overcurrent devices are described in Part II of Article 240. Overcurrent devices "must not be positioned above steps of a staircase," states Section 240.24(F). A major safety issue is addressed in this criterion. If a panelboard or switchboard is located in a stairwell, nobody ought to be forced to stand on different steps of the stairs to replace, operate, construct, troubleshoot, or restore overcurrent devices.

10. Answer: b

The particular rules for room air conditioners that are (1) cord-and-attachment plug connected and (2) not more than 250 volts ac, single-phase, are outlined in Part VII of Article 440. The overall rating of the air conditioner must be at most 50% of the rating of a branch circuit that also supplies lighting units, or appliances to avoid overloading the circuit in violation of Section 40.62(C).

11. Answer: d

The specifications for electrical systems' bonding are outlined in Part V of Article 250. Metal water pipe systems must be bonded to either the service entry enclosure, the grounded conductor at the service, or the grounding electrode conductor, according to Section 250.104(A)(1).

A breakdown of electrical insulation may cause the associated pipe systems to activate when mechanical and electrical connections are found inside the equipment. For instance, if the insulation fails, a water heater's electrical circuit may activate metal water pipes.

Metal water pipe systems need to be bonded to reduce the danger of electrical shock if a fault develops between an ungrounded (hot) conductor and the piping.

Also, remember that any metal pipe systems, including gas piping, that will likely become electrified must be linked to a grounded source wherever they are placed or attached to a building or structure. [250.104(B)]

12. Answer: a

The guidelines in Part II of Article 680 apply to the setting up of wiring techniques and apparatus around or within permanently erected swimming pools.

According to the specifications stated in Section 680.22(A)(1), each permanently installed swimming pool must have at least one (1) single-phase, 15- or 20-ampere, 125-volt outlet that is positioned between six and twenty feet away from the pool.

This regulation eliminates the need for extension cables near the pool while allowing common equipment, like a radio, to be securely plugged in and operated there. Please note that GFCI protection must be included for the receptacles. [680.22(A)(4)]

13. Answer: d

In Part II of Article 320, the installation requirements for armored cable, Type AC, are covered. According to Section 320.23(B), no guard strips or running boards are necessary to safeguard the cable in accessible attics when it is parallel to the sides of studs, beams, or floor joists.

According to Section 320.23(A), shielding of the cable is necessary if it is within six (6) feet of the closest edge of the scuttle-hole or attic entry if the cables cross the top of the floor joists in an attic that is not accessible by stairs or studs

Conclusion

Journeyman electricians must have accumulated 8,000 hours of work experience throughout their apprenticeship, typically lasting for four years.

Deciphering mechanical drawings and blueprints are one of the skills that aspiring journeymen are taught; put on your safety gear and ensure you're grounded; comply with all applicable electrical laws and regulations.

Even while journeymen and master electricians begin their careers with the same level of education, there are significant differences between them as their careers progress. Since master electricians are required to have a deeper understanding of a larger variety of electrical concepts, journeymen need more training before taking on greater responsibilities.

Wages are heavily influenced by factors such as geography. For instance, the salary of an electrician in Florida is much lower than that of an electrician in Oregon. This is mostly attributable to the fact that earnings, regardless of industry, closely reflect rises in the area's cost of living.

Illinois, New York, and Alaska are among the states with the highest average salaries for electricians. On the other hand, states in the South are known for having more stringent anti-union regulations. Perform some research before choosing; aspiring electricians should consider the surrounding region.

A journeyman electrician typically entails installing and repairing electrical cables and fixtures, installing electrical components, replacing or modernizing antiquated electrical systems, and installing and testing various pieces of electrical equipment.

Even if it's true that certain journeyman electricians are allowed to work freely under specific conditions, the rules in your area can be quite different from those in other places. Researching the local labor rules is essential to understand the tasks that journeymen may and cannot perform.

Glossary

A

AC (Alternating Current) — Alternating Current is referred to as AC. It is an electric current that regularly flips direction numerous times per second.

AFCI (Arc Fault Circuit Interrupter) – A specific kind of outlet or circuit breaker called an arc fault circuit interrupter opens the circuit when it detects a potentially hazardous electrical arc. It is used to stop electrical fires.

Apparent Power — Volt-amperes (VA). It is the sum of the rms voltage and rms current that yields apparent power.

Ampere (A) — It is the way to express how strongly an electric current flows across a circuit. A current flow of one coulomb per second corresponds to one ampere.

Admittance (Ω Ohms) — Resistance is fundamentally the antithesis of admission and is given by 1 divided by the resistance. It is a measurement of the amount of current that a device or circuit will let flow.

Ammeter — measures the amperes of current flowing across a circuit. In the circuit, an ammeter is linked in series, unless using a clamp meter.

Amp Meter— It is a tool for calculating the amperes per unit of electrical current flow.

Amplitude— It is the highest value of an electrical current pulse or wave.

Amplifier— It is a tool for raising voltage and power.

Aluminum Wiring— It is a form of electrical wire that was widely utilized between the mid–1960s and the late 1970s but is currently only employed in a very limited number of applications.

Arc Fault Protection— It safeguards the whole branch circuit against arc faults in both series and parallel.

Arc— It is the spark or electrical discharge that results from an electrical current shunting over an electrode gap.

C

Capacitor – It is a passive electrical device with two terminals used to store electrical energy for short periods within an electric field.

Circuit — It is a closed route via which current or voltage-generating electrons move. Circuits may be set up in series, parallel, or any of these configurations.

Circuit Breaker — It is a mechanism that automatically interrupts the flow of current in an electric circuit. After addressing the overload or failure's underlying cause, the circuit breaker has to be reset and closed to resume operation.

Conductor — It is any substance through which electric current may readily travel. Metals and other conductive materials have comparatively low resistance. The most prevalent conductors in the electrical industry are copper and aluminum wire.

Current (I) — It refers to the electric current passing via a conductor; the water movement via a pipe may be likened to an electric current. It is based on amperage.

Copper — It is a kind of electrical wiring that is now considered the norm in most contemporary applications.

Commercial Electrician — It refers to an electrician who works on commercial buildings' electrical systems.

D

DMM – A digital multimeter, often known as a DMM, is a device for measuring voltage, current, resistance, capacitance, temperature, and frequency in electronic systems. Get familiar with using a digital multimeter.

Diode — A diode is a semiconductor device having two terminals that usually only permit one direction of current flow.

DC (Direct Current) — Direct Current is referred to as DC. Electric Current, known as DC, only flows in one direction.

Demand — It is the power or related quantity's mean value over a certain period.

Dead— It means free from any electrical charge or connection.

E

Electrician— It refers to a tradesperson with a particular certification class who is authorized to install, maintain, or repair electrical wiring and electrical components for buildings, machinery, powerlines, and other equipment.

Electrical Apprentice — It refers to a person studying to become an electrician who is sponsored by an electrical contractor who offers on-the-job training and experience, and they are registered as students with a trade apprenticeship authority.

Electrical Plug — It is the part of an electrical cable that goes into the outlet or receptacle.

Extension Cord — An electrical cable may temporarily power an item when it is too far away from a receptacle.

Electrical Insulation — Electrical insulation is the conductor and wiring's dielectric coating.

Electrical Panel — It is a service box with circuit breakers that distribute electricity to circuits all around the building.

Electrical Contractor — It refers to an electrician legally permitted to work on electrical projects relating to planning, installing, and maintaining electrical systems.

Electrical Apprenticeship — A four-year training program comprises four levels of theory and tests, 6-8,000 hours of supervised work experience, and is required to become a licensed electrician.

Electrical Code — Regulations are amended or changed every three years, and electrical systems are installed and repaired.

F

Farad — It is a capacitance measurement unit. One coulomb per volt is equivalent to one farad.

Frequency — How many cycles occur each second in Hertz units? One cycle per second is considered one hertz (Hz); 60 cycles per second is considered one hertz (Hz).

Feeder — It refers to all circuit wires between the final branch-circuit overcurrent device, the source of a derived system, or other power supply source.

Fuse — A circuit interrupter is made of a wire strip that melts and interrupts an electric circuit when the current is too high. After addressing the failure's root cause, the fuse must be replaced with an identical fuse of the same size and rating to resume operation.

Fuse Panel — It is a service box with fuses inside of it.

G

Ground or Earth — A common return channel for electric current, a physical link to the Earth, or a reference point in an electrical circuit from which voltages are recorded.

Grounded Conductor – It is a conductor in a system or circuit that has been purposefully grounded.

Ground Fault – An unintended electrical connection between an electrical circuit's ungrounded conductor and the typically non-current-carrying conductors, metallic raceways, or earth is known as a ground fault.

Generator – It is a piece of hardware connected to an external circuit that will transform mechanical energy into electrical energy. There is a vast range of possibilities for the origin of mechanical energy, from a simple hand crank to a complex internal combustion engine. Most of the electricity that feeds into electric power systems comes from generators.

GFCI Protection — The ground fault circuit interrupter, often known as GFCI protection, is a safety feature that cuts off the flow of electricity to an outlet, preventing the risk of electrocution.

Grounded — It refers to anything in contact with an electrical current as well as the ground at the same time.

H

Henry — It is one of the units used to measure inductance. The inductance of a circuit is expressed in henry.

Hertz — It is the measure of how often something occurs. The old phrase cycle per second has been replaced by this (cps).

Handyman — A person who is not permitted by law to undertake regulated electrical work but who can carry out other common home duties and repairs

I

Inductance (H) — The characteristic of a conductor is that a voltage (electromotive force) is induced (created) in both the conductor itself and in any other conductors that are near it (mutual inductance) and calculated using Henries (H).

Inductor— An inductor is a coil of wire wrapped around an iron core.

Inverter — It is a piece of equipment capable of transforming direct electricity into alternating current.

Insulator — It refers to any substance that prevents an electric current from flowing freely through it. Insulating materials have reasonably strong resistance. Insulators prevent electric shocks from harming both equipment and people.

Industrial Electrician — An industrial electrician works mainly in commercial or industrial settings to complete electrical tasks.

Imaginary Impedance — In alternating current (AC) circuits, a "real" resistive component, which may also be present in direct current (DC) systems, has two additional hindering mechanisms known as reactive or imaginary components. The first is connected to the ever-shifting magnetic field and self-inductance, while the second is associated with electrostatic storage and capacitance.

K

Kilowatt-Hour (kWh) — It is the result of multiplying power, measured in kW, by time, in hours, equivalent to one thousand Watt-hours. For instance, the usage of a light bulb with 100W for 4 hours would result in the consumption of 0.4 kWh of energy (100W multiplied by 1kW divided by 1000W multiplied by 4 hours). Kilowatt-hours, or kWh, are the measure used to price units of electrical energy.

Kilowatt-Hour Meter — It is a gadget to measure the use of electrical energy.

Kilowatt (kW) — It is equal to 1000 watts.

Knockout Set — Knockout punches and electrical knockout sets are more precise names for this tool. When creating new holes in an electrical box or panel, an electrician's go-to tool is a knockout punch. A knockout punch set provides you with various knockouts in various sizes to choose from. Traditional knockout punches used by hand are tightened and loosened using a socket wrench.

Kelvin — It is the examination and evaluation of the color temperature that may be discovered in the light spectrum.

Knob-and-Tube Wiring — It is an early standardized kind of ungrounded electrical wire often utilized in buildings throughout the United States and Canada from the 1880s through the 1940s.

L

Load — It refers to anything that uses electrical energy, such as lights, transformers, heaters, and electric motors, and is considered an electrical appliance.

Lenz Law — Although Lenz Law is on the more technical end of things, one of the electrical engineers you deal with could bring it up. The direction of the current induced in a conductor by a variable magnetic field causes the magnetic field created to be in opposition to the initial magnetic field. It happens because the induced current produces a magnetic field that is proportional to the magnitude of the induced current.

Leakage — The term "leakage" refers to electrical current leaking from the circuit.

Licensed, Bonded & Insured Company — A licensed, bonded, and insured firm refers to a contracting company that has obtained the appropriate certification and insurance for the benefit and protection of its customers. Licensed, Bonded, and Insured Companies are Preferred Contractors with the Better Business Bureau.

N

Neutral Conductor — The conductor of a system is linked to the neutral point and has the function of carrying current when all other circumstances are typical.

O

Overload — The operation of equipment at a rating higher than the typical full-load rating or the operation of a conductor at an ampacity higher than the rated ampacity, either of which, when continued for an adequate amount of time, would produce damage or hazardous overheating. An issue, such as a ground fault or short circuit, is not the same as an overload.

Output — The current, voltage, or power a device or circuit provides is its output.

Ohm — It is a unit of measurement used to characterize the material's resistance to an electrical current.

Ohm Meter — It is a piece of apparatus that determines the number of ohms in an electrical current.

Open Circuit — When a circuit is broken, the current passage through the circuit is stopped, and the circuit is no longer functional.

P

Power — It is the speed at which an electric circuit may transport electrical energy from one place to another, referred in Watts.

Parallel Circuit — A circuit that has many different channels via which electricity may flow. Each load linked in its independent branch gets the whole voltage provided by the circuit, and the total current supplied by the circuit is the sum that results from the individual branch currents.

R

Rectifier — A piece of electrical equipment that, when connected to an alternating current, may change it into a direct current by only allowing electricity to travel in one direction via it. Both half-wave rectifiers and full-wave rectifiers may be found in use today.

Reactive Power — The fraction of an AC device's total electrical current responsible for establishing and maintaining the electric and magnetic fields is referred to as reactive power. There is a phase shift in an AC circuit if the current and voltage are out of sync with one another. It is calculated using VARS.

Resistance (Ω Ohms) — It is the resistance presented to the flow of an electric current. The resistance that electrical current encounters may be likened to the friction that water feels when it moves through a conduit, presented as a value in ohms.

Residential Electrician — An electrician who specializes in residential work is known as a residential electrician. This kind of electrician maintains and repairs electrical systems in homes.

Receptacle — A receptacle is an electrical outlet that allows for the connection of various devices.

Rewiring — Rewiring refers to replacing the wiring already present in a structure, appliance, or device with new wires.

S

Series Circuit — A single-ended circuit is a circuit with only one possible route for electricity to flow. For the circuit to be complete, the whole current must go through all the loads. Only then can it return to its original supply.

Stripping — It means taking off the insulation that covers a conductor or wire.

Series Parallel Circuit — An electric current contains groups of receptive devices linked in parallel, while the groups themselves are grouped in series inside the circuit, also called a series of multiple circuits.

Short Circuit — A defect in an electric circuit or device, often caused by inadequate insulation, results in the flow of current taking another route, which either causes harm or causes the current to be squandered.

Service — The wires and other equipment are responsible for transporting electrical current from the electrical supply system to the system being serviced.

Service Lateral — The underground service conductors are located between the street main, which includes risers, and the initial point of connection to the service-entrance conductors in a terminal box.

Semiconductor — A solid material that, due to the presence of impurities or as a result of temperature, has a conductivity that can be placed between that of an insulator and that of the vast majority of metals. The majority of electronic circuits need components fabricated using semiconductors, specifically silicon.

SCR (Solid State Relay) — An electrical switching device can be turned on or off by applying a tiny amount of external voltage across its control terminals, causing it to switch. The activity of switching takes place in a very quick manner.

Solenoid — When an electric current flows through a spiral of conducting wire, its turns become virtually equal to a sequence of parallel circuits. The wire takes on magnetic characteristics comparable to those of a bar magnet.

Switch — The ability to make, break, or otherwise alter the connections that make up an electric current is the primary function of a switch.

Switchgear – Fuses, circuit breakers, and electrical disconnect switches are used to regulate, safeguard, and isolate electrical equipment. Switchgear has two primary purposes: it de-energizes equipment so that maintenance may be performed on it and clears errors that have occurred further down the line.

Substation — It refers to a group of electrical components that work together to increase, decrease, or otherwise control the voltages of electrical currents.

Surge Protector — This is a piece of equipment intended to shield electrical appliances and other devices from sudden surges in voltage.

T

Transistor — It is a component made of semiconductor material with three connections that can rectify and amplify signals.

True Power — Referred to in Watts, it is the power materialized in observable forms such as acoustic waves, electromagnetic radiation, or mechanical events. Voltage and current flow in the same direction in a circuit using direct current (DC) or alternating current (AC), provided the circuit impedance is pure resistance.

Transformer — It is a piece of electrical equipment that may alter the voltage of an electric current.

V

VARS — It is one of the units used to quantify reactive power. When the voltage and current are stated in volts and amperes, vars may be considered the imaginary component of visible power or the power flowing into a reactive load. Both of these interpretations are valid.

Voltage Drops — It is the decrease in pressure brought on the conductor resistance in an electrical circuit.

Volt-Ampere (VA) — Perceived power is measured by multiplying the rms current by the rms voltage.

Volt (V) — It is a voltage measurement unit. The difference in potential required to force one ampere of current through a one-ohm resistance is equivalent to one volt.

Voltage (E) — Like water, pressure causes water to flow through a conduit; an electromotive force or "pressure" causes electrons to move. It is expressed in volts.

Volt Meter — It is a tool for detecting an electrical current's voltage.

W

Watt-Hour (Wh) — A measure of electrical energy equals one watt of electricity used for one hour.

Watt (W) — It is an electrical power metric. In an electric circuit, when there is a one-volt potential difference and a one ampere current, one watt is equivalent to one joule per second, or the power.

Watt Meter — A watt meter is a tool used to calculate the electrical power consumption of a circuit in watts.

Made in the USA
Las Vegas, NV
24 November 2024

12525411R00107